CAUTION !
Reading This Book
Can Make You Think

OTHER BOOKS BY THE AUTHOR

- *Captain Bucko's Nauti-Words Handbook*, Fascinating Facts and Fables About the Origins of Hundreds of Nautical Terms and Everyday Expressions, (ISBN 0-595-31529-1).

- *Fresh Earthworms Taste Green (The Early Years)*, Entertaining Short Stories of a Typical Kid growing up in Mid-America Among Tombstones, Indian Bones, Pet Skunks, Rhythm Bands, Vindictive Parrots, UFOs, Gigantic Chickens, Love Slaves, and Lollipops, (ISBN 0-595-34385-6).

- *Captain Bucko's Water & Weather Handbook*, An Entertaining and Easy-to-Read Collection of Inside Information, Fascinating Facts, Trivial Tidbits, and Helpful Hints by a Professional Oceanographer and Marine Meteorologist, to Help Make Your Voyages Safer and More Enjoyable, (ISBN 0-595-39687-9).

- *Fresh Earthworms Taste Green (The Middle Years)*, More Entertaining Short Stories of a Typical Middle-Class Kid growing up in Middle America during the Middle of the 20th Century, despite the Challenges of Aquatic Archery, Attempted Socialization, Haunting Caves, Embarrassing Dates, Odd Summertime Jobs, Cruising Streets, Nocturnal Skiing, Being a Gorilla, Pirate, & Rock Star, plus Other Great Adventures, (ISBN 976-0-595-49489-7).

- *Captain Bucko's Galley Slave Cookbook,* Fascinating Facts, Sea Stories, & 100+ Famous Recipes From Worldwide Ports of Call, (ISBN 978-0-595-44537-0).

- *Journey of the Lost Princess,* Adventure and Romance in the Mysterious Land of the Incas, (ISBN 978-1-45024-305-6).

As a former Army Brat and a Navy Veteran, Roger is familiar with the sacrifices routinely made by our military service members and their families. In recognition, he is donating the profits from *Journey of the Lost Princess* to top-rated charities that support these real American heroes. Copies of Roger's books are available from all good bookstores, or ordered online at *www.BarnesandNoble.com* and *www.Amazon.com*. For additional information about his work, please visit his Website at www.WriteAweigh.com.

CAUTION !
Reading This Book
Can Make You Think

The Premier Mental Exercise Program
for Flabby Modern Minds

Roger Huff

iUniverse LLC
Bloomington

CAUTION ! READING THIS BOOK CAN MAKE YOU THINK
The Premier Mental Exercise Program for Flabby Modern Minds

iUniverse books may be ordered through booksellers or by contacting:

iUniverse LLC
1663 Liberty Drive
Bloomington, IN 47403
www.iuniverse.com
1-800-Authors (1-800-288-4677)

ISBN: 978-1-4917-1716-5 (sc)
ISBN: 978-1-4917-1717-2 (e)

Library of Congress Control Number: 2013922206

Printed in the United States of America.

iUniverse rev. date: 12/12/2013

Contents

To Military Members and Their Families,
Our True American Heroes

ACKNOWLEDGEMENTS

The author thanks his family and friends for their proofreading skills, constructive suggestions, and continued support; and acknowledges other authors, historians, and researchers for their contributions toward providing a balanced perspective about controversial topics. The author also wishes to express his gratitude to all past and present military service members and their families for their sacrifices that earn the freedoms to publish books like this. Thank you for your inspiration, patriotism, service, courage, and understanding.

INTRODUCTION

This journey began many decades ago and led more than halfway around the world. As a curious kid, I destroyed a lot of my toys in trying to figure out how they worked. My parents let me make mistakes, learn from experiences, and somehow tolerated my early experiments and my playing of rock & roll music. Working with sharks and searching for shipwrecks in the Bahamas and Florida Keys met the need for adventure during my college years; and after completing Master's programs in Oceanography and Meteorology, I served the next twenty years as a Navy Geophysics Officer.

It took me from the Mediterranean to Micronesia and rekindled my long-term interests in learning how very different and lesser-known societies evolved and survived. During a subsequent Information Technology career in the corporate world, I noticed how politics had begun to influence science and how unethical people take advantage of those who have lost the ability (or desire) to think for themselves. These experiences and concerns led me to undertake this project.

This is a very different kind of book. It explores an extremely broad range of tantalizing topics and intentionally interleaves its challenging Chapters with humorous ones. It will make you smarter, a witty conversationalist, a lot more attractive, a better athlete, and (possibly) able to leap tall buildings in a single bound. Its varying subjects, unique format, and tongue-in-cheek humor are all cleverly crafted to capture and hold your interest, allowing you to sharpen the skills required to survive in modern society.

The survival of primitive societies depended upon the skills of their members to avoid being eaten or stepped on by large beasts. As the number of such beasts dwindled, the threats they posed were replaced by ones that arose from within societies themselves; and the most serious threats faced by modern societies include: self-centered greed, naïve ignorance, and irresponsible apathy.

Self-centered greed destroys the very mortar that holds societies together. Its seeds are sown when a society's members begin to lose respect for moral and ethical values; and it grows when they become more concerned with what they legally *can* do than what they morally *should* do. Greed has been around for millennia; and it has caused a number of great societies to crumble.

> *"There is a sufficiency in the world for man's need, but not for man's greed."*
> —Mahatma Gandhi

Naïve ignorance breeds in brains that become content with being spoon-fed information by a biased media or others attempting to control public opinion. Symptoms include the inability to: tell facts from fiction, learn lessons from history, communicate effectively, or independently think clearly.

> *"Nothing in the world is more dangerous than sincere ignorance and conscientious stupidity."*
> —Dr. Martin Luther King Jr.

Irresponsible apathy is the most insidious of the three because it often spreads undetected throughout societies until it is extremely difficult to overcome. It starts when otherwise good people begin to foolishly believe that their society's survival is not their responsibility, and are simply too lazy to challenge what they hear or read. Unchecked apathy lets self-centered greed and naïve ignorance prosper and prevail, and that is typically a terminal condition.

"The world is a dangerous place to live; not because of the people who are evil, but because of the people who do not do anything about it."
—Albert Einstein

These threats are just as real and dangerous as ones posed by hungry, clumsy beasts; but ensuring survival of a modern society is considerably more complex than simply stockpiling bottled water, freeze-dried food, and ammo in bunkers buried on the outskirts of places like West Bugshuffle, Idaho. As societies grow larger, their members typically become more interdependent; and the skills of fewer individuals have more effect on the survival of their society as a whole.

Advanced societies have not survived catastrophic events merely because there were not enough of their members with critical skills left to rebuild them. These skills have migrated to the opposite end of our bodies; and surviving the threats to modern societies requires more brains than brawn. There are plenty of books about physical exercise, but no modern society has ever survived solely because its members had rock-hard abs or low body-fat ratios.

This is a unique exercise program for flabby modern minds that applies the same methods proven in physical fitness routines. It starts with simple warm up exercises to prevent brain strain, and varies subjects to avoid mental monotony. To be more effective, some of its exercise elements are intentionally designed to require extra concentration; and its seemingly silly sections provide "cool down periods" between more controversial or difficult Chapters.

Both types are important because topics that appear to be arbitrarily chosen are actually subtly-structured building blocks in the larger program to develop critical skills; and humor is used to make learning more fun and effective by associating newly-gained knowledge with a silly saying, giggle, or guffaw.

Skipping over a challenging Chapter would also deprive the reader of those exercise benefits.

Sharpening the skills necessary to survive self-centered greed, naïve ignorance, and irresponsible apathy will require: *curiosity* about unfamiliar subjects, *courage* to question information, and *commitment* to make the effort to learn more. So, do you have what it takes?

WARNING!—Over-exercise is counterproductive. To avoid mental meltdowns, **don't** try to read this entire book at one sitting. On the other hand, **do** read the *Exercise Guides* near the beginning of each Chapter that suggest the best ways to approach its exercise elements. To achieve optimum exercise benefits, take brief *pauses* and longer *breaks* at the points indicated with giant asterisks (✴) to review material and let your brain expand to accommodate all your newly-gained knowledge. With these helpful hints in mind, you are ready to begin with a few . . .

WARM UP EXERCISES

Couch potatoes don't just wake up and decide to compete in a triathlon that day. Over the six to nine months it typically takes to prepare, they must learn how to swim, ride a bike, and run; give up all the things they really enjoy eating, have a lobotomy, hire an expensive personal trainer, and focus on developing muscles below their necks. The morning of the event, they need to stretch and indulge in light warm-up exercises like a 200 to 500-yard swim, a 15-minute bicycle ride, and a 20-minute run. Once they have essentially re-directed all of their blood away from their brains, they are ready to compete.

Preparing to enjoy this book is quite similar, except for all the many differences. You don't need to learn how to swim, bike, run, or even get up off of the couch. You don't need an expensive personal trainer, a lobotomy, or to give up any of your favorite foods. You don't need to develop any muscles at all, but you do need to warm up your out-of-shape brain a little bit before starting.

Exercise Guide: This introductory section includes four warm up questions that people have pondered for years but were too timid to ask aloud. I recommend that you ask yourself each question prior to reading the provided answer, and pause briefly afterwards to let the information sink in. You are obviously much smarter than most simply because you are reading this book, but I shall begin easy and promise to be gentle. Here's your first mental warm up question:

ROGER HUFF

(How) Do Whales Sleep?

As a marine biologist, I was asked this question a number of times, along with ones like: "If whales sleep, how do they breathe?" and "If whales breathe, how do they keep from drowning?" The answers are fairly fascinating, so keep reading.

Unlike humans whales are voluntary breathers, which means they have to think about each breath they take. And since they breathe through blowholes atop their heads, they must also come to the surface to do so. So how can whales ever get any sleep?

Whales actually "sleep" by resting half of their brain at a time while the other half remains awake to make sure they breathe properly and stay alert for danger. They also don't appear to exhibit any Rapid Eye Movement (REM) sleep, the stage during which most dreaming occurs in humans; and their breathing becomes more intermittent than when they are active.

Where they choose to sleep varies among species. Some, like humpbacks, rest on the surface and remain so motionless it's called "logging," because they look like giant floating logs. Other whales constantly swim while they're asleep, while captive dolphins may rest at the bottom of their tank for several minutes.

✻ **Pause briefly to reflect upon those snoozing leviathans.**

Why Are There Rainbows?

When people are asked this question, they often come up with answers that are strange (although creative) combinations about dreams, making wishes, physics, the Wizard of Oz, weather proverbs, pots of gold, and leprechauns. Armed with the following explanation by our National Weather Service (NWS), however, you can flabbergast folks with incredible intelligence and baffle them with brilliance.

Rainbows are some of the most spectacular phenomena observed on the Earth. They are formed when visible light (typically from the Sun) strikes more or less spherical water droplets in the air. When it enters these water droplets, light is first refracted (bent), reflected (at least once) from the droplet's internal curved surface, then refracted once again as it leaves the droplet and re-enters the air. But the various colors (wavelengths) in the visible light spectrum are affected differently by these processes, creating the optical effects we call rainbows.

Rainbows do not necessarily involve rain, but can also be produced by spray, mist, or even airborne dew. In so-called "primary rainbows" created by a single reflection of light within water droplets, the longer wavelength colors like reds, oranges, and yellows appear on the outer edge, while the shorter wavelength colors like indigo, violet, and blue are on the inner edge. Monochromatic (i.e., single color) rainbows occasionally happen at sunrise or sunsets, when most shorter wavelengths are no longer present; and so-called "moonbows," which usually appear as shades of white, are sometimes seen in bright moonlight.

If light reflects back and forth more than once before leaving the water droplets, multiple rainbows can result. Primary and secondary rainbows may be observed when:

- The light source (e.g., the Sun) is at the observer's back and the water droplets are in front, and

- The Sun is at an angle of forty-two degrees or less above the horizon.

The colors in secondary rainbows are in reverse order of those in primary ones and are typically also fainter. Rainbows are actually circular, but we usually only see portions of them unless we're looking down on them from an airplane or high mountain. And since they are only optical effects, you cannot ever actually reach a rainbow's end and find that pot of gold. Those wily little leprechauns never told you this, did they?

✳ Pause now and review what you learned about rainbows.

Has your brain warmed up yet? Does it need a little more exercise? If so, here is a third question to stimulate that flabby gray matter:

Where Did The Easter Bunny Originate?

If you already know the correct answer to this you may skip to the next question, but if not you had better keep reading. The origins of many holiday symbols are fairly well known: decorated trees for Christmas, jack-o-lanterns for Halloween, menorahs for Hanukah, and Uncle Sam on Independence Day. But how did a rabbit and eggs ever become associated with Easter?

The relationship actually dates back centuries, before the holiday was associated with resurrection of Christ. Long before 325AD, when the Council of Nicaea ruled that the Christian holiday was to be celebrated on Sunday, pagans held festivals honoring *Eostra* (or *Ôstarâ* in German), goddess of spring and fertility, on various days at the time of the Vernal Equinox. To ease pagan conversion to Christianity, missionaries co-opted their holiday, and it eventually became known as Easter.

But what does this have to do with rabbits? As it turns out, hares and rabbits were old Eostra's earthly symbols, because they were the most fertile animals known. The Easter Hare was first mentioned in German literature in the 1500s, and the symbol immigrated to America with the settlers in Pennsylvania Dutch country during the 1700s. Children would construct tiny nests in their caps and bonnets, and hope that *der Osterhase* (The Easter Hare) would fill them with colorful eggs.

Eggs were also a symbol of renewal and fertility, and these early ones were just shells after the insides had been blown out and used for cooking. But the Easter Bunny wasn't widely accepted outside German Christian communities until after the Civil War, because religious leaders did not like associating a pagan symbol with Easter. And by the way, the first edible Easter Bunnies were made out of sugar and pastry, not chocolate.

✳ **Pause to absorb the above information before hopping off.**

Now when you hear old Gene Autry sing, "Here Comes Peter Cottontail," it will mean more to you. So, are you ready for your final mental warm-up question?

How Was the Moon Formed?

Our Moon's origin has been the subject of much speculation for centuries, but the prevailing theories weren't supported by analyses of lunar rocks collected during the Apollo missions, which began in the late 1960s. During the late 1970s, a new theory suggested that the Moon formed after a gigantic glancing collision roughly 4.5 billion years ago between a nearly-formed "Protoearth" and a planet having a mass roughly the same as Mars. The energy released during the impact melted the Protoearth's outer crust; and the debris from both bodies that was cast into space subsequently coalesced into the Moon. This so-called "Big Whack" theory is currently a leading contender, but don't go away because things may change.

The Moon is in synchronous rotation with the Earth, which means we always see its same side. Its gravitational attraction influences our tides, helps stabilize the Earth's rotational axis, and causes a slight lengthening of our days. But the Moon is moving away at about 1.5 inches a year, so you had better enjoy it now.

✳ **Take a break at this point, because astrophysics can be stressful.**

That's enough mental warm up! You are now be ready to begin examining even more interesting and challenging subjects, and we will start by

CHAPTER ONE:
GETTING REAL

This Chapter lays the foundation for subsequent exercises by focusing upon the ability to distinguish rumors from realities. People who lack this fundamental skill tend to believe whatever they read or hear, which can lead to naive ignorance and terminal stupidity. Questioning popularly-held opinions sometimes takes a little courage, but remember:

> *"No matter how big the lie; repeat it often enough and the masses will regard it as the truth."*
> — John F. Kennedy

Exercise Guide: You should have no problem completing this relatively short Chapter in a single sitting. However, pausing briefly and taking slightly longer breaks at the points indicated by giant asterisks (✱) to review and digest what you have learned are both highly recommended.

There are a lot of myths, biases, opinions, hypotheses, and theories out there scattered among relatively few facts; and it can be a major challenge to identify them correctly. To begin, let's review the official definitions for each of them from the Merriam-Webster Dictionary:

> ***Myth***: *A usually traditional story or parable of ostensibly historical events that serves to unfold part of the world view of a people or explain a belief, a practice, or a natural phenomenon. A popular belief or tradition that has grown*

up around something or someone, especially embodying the ideals and institutions of a society or a segment of society. A person or a thing having only an imaginary or unverifiable existence.

One example of a popular myth is: "Witches were burned at the stake in Salem Town, Massachusetts." Actually, about 150 people were arrested during these infamous trials, 31 were tried, and 20 were executed; but none were burned at the stake. Hanging was the most commonly used method, but at least one was pressed to death beneath heavy stones.

Myths sometimes become realities. For example, sailors' tales of sea monsters were considered to be myths for centuries, until the first confirmed carcass of a Giant Squid was recovered during the late 1800s. Other formerly mythological beasts include the Okapi, and my personal favorite, the Duckbill Platypus. But, have you heard the latest:

Rumor: *Talk or opinion widely disseminated with no discernible source. A statement or report currently without known authority for its truth. Talk or report about a notable person or event.*

Notice the official definition does *not* imply whether or not a rumor might be true, but rather that it's currently without a discernible source or proving authority. One of the furthest-reaching rumors recently was Y2K, or the "Millennium Bug," which claimed due to a glitch in the way computer programmers had encoded dates at the stroke of midnight on December 31, 1999, modern societies would come to a screeching halt, airplanes would fall from the sky, our communications systems and power grids would fail, bank records would be lost forever, and McDonalds would simultaneously run out of Chicken McNuggets. None of those occurred.

✳ Pause and review the information about *Myths* and *Rumors*.

Many myths and rumors are debunked or simply fade away, while others evolve into popular opinions. If and how the latter occurs sometimes depends upon the prevailing:

> **Bias**: *An inclination of temperament, especially personal and sometimes unreasoned judgment. An error introduced into sampling or testing by selecting or encouraging one outcome or answer over others.*

Notice that the official definition of bias includes the word "sometimes," which means this common trait can be based on reasoned principles. It is important to keep in mind that bias is distinctly different from prejudice and bigotry, and bias can result in positive as well as negative decisions. Well-founded biases help us all make decisions about: how we live, what we eat, what is dangerous, and with whom we choose to associate. And bias can and does also influence our:

> **Opinion**: *A viewpoint, judgment, or appraisal formed in the mind about a particular matter. A belief that is stronger than an impression, but is less strong than positive knowledge.*

We have lots of opinions. There are legal ones, editorial ones, political ones, and personal ones. There are public opinions, ethical opinions, ones about where to get the best pizza, and opinions regarding the existence of Unidentified Flying Objects (UFOs). Here's one of the latter now:

It was 1952, and my family was attending an ice cream social at Webster School. I was near the swing set with my classmate Henry Sullivan, when we saw a very bright object pass rapidly and silently across the evening sky. This was five years before Sputnik, and it was faster and quieter than any plane of the day. We could not identify it, it was obviously flying, and it was clearly some sort of object; so by definition it was a UFO. My opinion is that UFOs exist because I saw one, but as far as their extraterrestrial origin, I might need a little more convincing.

Unless you happen to be a spineless jellyfish or a politician running for election, there is nothing inherently wrong with having well-founded bias or opinions. But as you will learn in subsequent sections, when prevailing bias is powerful and unchecked, it can result in unproven opinion becoming accepted as:

> *Fact: The quality of existing in act or reality and not merely potentially. being actual or real. Something that has actual existence. Information presented as having objective reality.*

✱ **Pause here to review the definitions of *Bias*, *Opinion*, and *Fact*.**

The preceding are perfectly valid terms, but are a little too subjective for some. Folks with extra letters after their names (e.g., PhD) prefer to use terms based upon ancient languages like Greek and Latin, so we also need to mention the following fancier ones:

> *Hypothesis: Essentially an "educated hunch" based upon observations, and which can be disproven or supported through additional observations or experimentation.*

For example, suppose you buy winning lottery tickets three weeks in a row and hypothesize, "If I buy more lottery tickets, I will become rich." Unfortunately, just having such a hypothesis does not affect the actual odds, which you learn after investing your life savings in worthless lottery tickets. And since it has not been thoroughly tested yet, a hypothesis may pose a weaker argument than a:

> *Theory*: *Scientific theories describe a hypothesis or group of hypotheses supported by repeated experimentation or testing. If sufficient evidence accumulates to favor hypotheses, they may become accepted as theories, and theories remain valid for as long as there is no credible evidence to dispute them.*

For example, most dinosaurs became extinct rather suddenly (about sixty-five million years ago, or at the end of the Cretaceous Period for those of you taking notes), and there are several scientific theories about just why this occurred. One of the most persistent evolved out of a 1980 hypothesis that it resulted from the effects of an asteroid impact off the Yucatan peninsula. Because there has been evidence from other scientific disciplines to support this theory, it remains valid.

✳ **Pause to review what you've learned about *Hypothesis* and *Theory*.**

Science & Philosophy

There are significant differences between *scientific* theories and *philosophical* ones. Scientific theories must be supported by results of objective experiments, testing, and/or independent research, and have to be accepted by the scientific community. Philosophical theories, on the other hand, may include empirical and non-empirical statements, are not necessarily testable scientifically, and merely have to be accepted by general segments of societies.

Philosophical theories are believed to be true by those who accept them as such, and form the bases for: schools of thought, ideologies, belief systems, organized religions, philosophical movements, and world views. The following "isms" are forms of philosophical theories: Altruism, Buddhism, Capitalism, Communism, Darwinism, Empiricism, Freudianism, Humanism, Idealism, Judaism, and Pessimism. There are plenty more, but these examples illustrate that philosophical theories cover an extremely broad spectrum.

Determining whether or not something qualifies to be a theory is obviously quite different in science and philosophy, and can depend on one's definition of what constitutes "proof." Scientific proof is defined as reaching a logical conclusion based on objective and overwhelming evidence; but in the philosophical world, evidence is often defined as "that which justifies belief."

When Theories Conflict

As long as neither is disproved, there is nothing to prevent contradictory theories from simultaneously existing. There are plenty of examples of conflicts between scientific and philosophical theories throughout history, and the following one illustrates how very powerful bias can affect what becomes accepted as fact.

People had already been wondering why the Sun, Moon, and stars appeared to move across the sky for millennia by the time the geographer, astronomer, and mathematician named Ptolemy suggested that it was because the Earth was the center of the Universe. And since this fit quite nicely with prevailing philosophies, his geocentric view became widely accepted for many centuries afterwards.

Sometime around 1512AD, an ecclesiastic, canon lawyer, physician, student of mathematics, and astronomical party-pooper named Copernicus suggested a heliocentric, or Sun-centered, concept. Parts of old Ptolemy's view had already been questioned by Islamic astronomers, but the primary reason geocentrism didn't quietly fade away was a great deal more philosophical than scientific.

In 1615, a Dominican Friar filed a complaint with the Inquisition against the Italian astronomer, mathematician, and physicist Galileo who supported the Copernicus view, because the church said it contradicted scripture. The church's Theological Advisory Committee concluded that the notion of Earth orbiting around the Sun was "a foolish and absurd mistake of faith," and was officially considered heresy.

Forced to recant his belief in heliocentrism, Galileo spent the rest of his life under house arrest and was ordered to read seven penitential psalms every week for the next three years. In 1983, the church finally admitted that Galileo might have been correct; and in 1992, Pope John Paul II expressed regret at how the church had handled the matter and acknowledged that errors were committed by the Committee in judging Galileo's views. It only took them 377 years to apologize.

Beliefs are quite personal, and no matter how sanctimonious we think we are, none of us is truly in a position to judge the beliefs of others. Analyzing treatises like the *Summa Theologica* is beyond the scope of this book, but that fact doesn't preclude applying logic and reason to philosophical theories. Faith is a keystone of many, but changes in the popularity or acceptance of philosophies and belief systems often depend more on their intrinsic merits perceived by a population.

Reality Check

You might wish to revisit this Chapter periodically to refresh your knowledge of the major differences among myths, rumors, opinions, facts, hypotheses, and theories. In it we noted that bias can be good or bad, provided an example of how prevailing bias can corrupt public opinion and affect what is accepted as fact, and examined fundamental differences between criteria for scientific and philosophical theories. With your skills at distinguishing rumors from realities honed to a razor edge, you are better prepared to fearlessly foray into a very controversial topic to further exercise your clear thinking skills. But first,

✻ **Take a break to review the information in this Chapter.**

CHAPTER TWO:
CLIMATE CHANGE

There is a tremendous amount of misinformation regarding this subject, and this Chapter provides advanced exercises in distinguishing among rumors, opinions, hypotheses, and theories. It is one of very few places today where realities about this controversial topic are objectively examined by a professional meteorologist and oceanographer with two decades of practical experience in environmental forecasting and atmospheric research, who is not biased by vested interests or promoting any particular group or agenda.

NOTE—This Chapter is not meant to be a complete treatise on climate change, but rather to exercise your skills at distinguishing rumors from realities and to examine how politics and ideologies have crept into its public perception.

Exercise Guide: To achieve the best exercise results, pause briefly to review and take longer breaks at the points indicated. With this approach, you should have little problem in completing this Chapter within a single day.

Changes in our planet's natural environment have played major roles in shaping human history, but climate change remains one of today's most misunderstood subjects. To begin, however, let's put "change" into proper perspective.

Change is a Constant

Can you name a single tangible thing beyond the atomic or molecular level that can be objectively proven has not changed since first appearing? Spoken and written communications have obviously changed; our beliefs and philosophies have clearly changed; Earth's dominant life forms have changed; the known Universe has changed; and so have human beings. Things that appear to be constant on one time scale are mere snapshots on another. Mountains and valleys are both on their way to becoming plains, and manmade structures all eventually crumble. Cultures come and go; while plans, priorities, and acceptable behaviors change; as do favorite pastimes and technologies.

Those of you who remain unconvinced probably don't remember: when families gathered around AM radios to listen to *The Lone Ranger's* latest episode, eight-track tape players, or Studebakers. The concept of change is so popular it has even crept into the vocabularies of our politicians, who promise change in their campaign speeches. The scale of changes they can actually deliver, however, pales compared to those routinely accomplished by Mother Nature. But before getting into that, let's explain the basic differences between:

Weather and Climate

I can understand confusing "Calvary" with "cavalry," saying "nuculer" instead of "nuclear," or pronouncing "asked" as "axed"; but as a meteorologist, it concerns me when I hear individuals who ought to know better use weather and climate incorrectly or interchangeably. There are very significant differences between these two terms, which is why we begin this Chapter with the following official definitions of each by the World Meteorological Organization (WMO):

> *Weather—The state of the atmosphere at a particular time, as defined by the various meteorological elements.*

The first important phrase in its official definition is "the state of the atmosphere," which means that weather describes the *actual physical conditions* existing within the envelope of gases surrounding our planet. The next important phrase is "at a particular time," which indicates the transitory nature of weather phenomena. The third important phrase in this official WMO definition is "various meteorological elements" which may include: humidity, temperature, atmospheric pressure, clouds, wind, precipitation, fog, etc.

Weather comes in many sizes and shapes, ranging from a small microburst just forty feet across to a Category Five hurricane hundreds of miles in diameter. It can last for only a few seconds or persist for several consecutive days or weeks. Even regional droughts and monsoons that last for months at a time are typically considered weather conditions, although they're sometimes mistaken for climate changes. Weather is essentially a snapshot of the actual atmospheric conditions that exist for a relatively brief period of time; and as you will see from the next official WMO definition, weather is definitely not the same thing as:

> *Climate—A synthesis of weather conditions within a given area that is characterized by long-term statistics (mean values, variances, probabilities of extreme values, etc.) of the meteorological elements in that area.*

An important term in this next official definition is "synthesis," which means that climate is actually a *derived*, instead of a naturally occurring, condition. Another important phrase is "within a given area," which indicates climate is actually a *local or regional* characteristic, not a global one. The third key phrase is "long-term statistics," which means that:

- Climate values should reflect valid data collected over many years or longer and

- As a computational result, climate values can be shaped by the input data accuracy, location, time frame, etc.

Because climate is a *synthesized mathematical product*, such a combination of meteorological elements may never actually exist with that particular area. It is not unusual for climate changes to indicate an increase in the average value of a specific parameter (e.g., temperature, precipitation) in some area(s), while the average value of the same parameter simultaneously decreases in others. Such details are quite important because they illustrate how climate changes depend upon the geographic area(s) and the time frame(s) chosen.

Professional gamblers know a thing or two about statistics, but since most of us won't admit to dealings with *Bernie the Bookie* on a regular basis we'll use golf and bowling for our analogy. Weather is similar to one's actual score made on a specific hole/frame, course/alley, and day. On the other hand, climate is similar to one's statistical handicap, which to be valid, must be computed over a much longer period of time. Changes in precipitation, temperature, or winds that persist for days, weeks, or even months are a lot more likely to reflect variations in large-scale *weather* patterns than in *climate*. The latter is statistical value that may never actually exist in the real world; and as golfers, bowlers, and bookies all know, changing long-term averages can be difficult.

✱ **Pause and review the above about *Change*, *Weather*, and *Climate*.**

Natural Equilibrium

The Universe, our Solar System, the Earth, and life itself are incredibly complex and interrelated balancing acts. They all respond to forces that none of us fully understand, very few of us ever think about, and many of us don't even know exist. They all endlessly strive for, but never fully achieve, states of equilibrium. Humans create theories and ideologies to explain things to each other, but none of us can be truly objective because we are all are parts of these processes.

Nature constantly attempts to maintain equilibrium even though most people are totally unaware this is even happening. It is what makes the wind blow and rivers flow downhill. This is what causes most earthquakes and the long-term changes in climate, whether humans have anything to do with them or not. And when we become too infatuated with ourselves, nature has ways to prove who is really in charge. Just ask the U.S. Army Corps of Engineers about controlling floods of the Mississippi River, or the crews of fishing boats on the Bering Sea, or the family in California whose house was destroyed by a massive earthquake, or the Colorado snowmobiler who got buried in an avalanche.

Temperature Change Triggers

Sometimes nature "overshoots" its goal of equilibrium, causing changes to swing back in the opposite direction. Folks who do not recognize this ongoing balancing act or are driven by biased agendas often interpret such reverses inappropriately, but temperature changes are certainly nothing new. During one extremely warm period (roughly 251 million years ago) an estimated 70 percent of all life on land and 80 percent of all ocean organisms perished, and in the last 650,000 years there have been a number of major temperature changes. The principal "triggers" for them include:

Volcanic Activity—which emits greenhouse gases like: Water Vapor, Carbon Dioxide (CO_2), Methane (CH_4), and Aerosols (e.g., ash, smoke) into the Earth's atmosphere. There is evidence that atmospheric CO_2 levels and temperatures have been considerably higher than they are today due to prehistoric volcanic activity. Conversely, the United States Geological Survey (USGS) has said that aerosols emitted by an Indonesian volcano in 1815AD blocked enough sunlight to *lower* global temperatures by as much as five degrees Fahrenheit (F).

Solar Energy Variations—The non-circular shape of our planet's orbit, periodic changes in the Earth's (22.1 to 24.5 degree) tilt toward/away from the Sun, and solar activity fluctuations (e.g., sun spots), all affect the amount of solar energy that reaches various parts of the globe. There is significant evidence that solar variations are a trigger for many temperature changes, and that reduced solar energy played a major role in the "Little Ice Age" between 1550 and 1850AD, when cooler temperatures occurred in North America, Europe, and elsewhere.

Ocean Current Changes—Because they are the principal mechanism for re-distributing heat energy throughout the globe, any heating, cooling, ice, salinity, or land mass changes that alter the ocean currents can trigger substantial temperature changes from region to region.

Greenhouse Gases—can affect heating and cooling of the Earth, but the reverse is also true. For example, if fluctuation in Earth's orbit and its associated change in solar energy cause the temperature to increase, more CO_2 will be released from the oceans, increasing the amount of that greenhouse gas. On the other hand, when the temperature decreases, more CO_2 becomes dissolved in the ocean, thus contributing to additional cooling.

Continental Movement—may not be mentioned as frequently as other factors, but when one is talking about changes over millions of years it can't be ignored. The reality is that most major continents are not in the same geographic locations they once were due to "plate tectonics," and this affects temperature changes.

Massive Energy Releases—Atmospheric explosions, or impacts with celestial objects like asteroids and meteors, as well as large-scale nuclear detonations, might also trigger temperature changes. In addition to the initial blast, thermal and relatively localized fallout effects, such events could at least theoretically propel enough particulate matter into the upper atmosphere to significantly decrease the amount of solar energy reaching the Earth's surface.

Large-Scale Changes

Depending upon the criteria used, there is substantial geological, chemical, and other evidence that Earth has experienced between three and eight wide spread temperature changes. All of them were caused by natural events, and they all occurred before the appearance of modern humans.

Changes come in a variety of shapes and sizes. At one end of the scale are the so-called "Ice Ages," periods of geologic time that typically span millions of years when:

- Temperatures decrease dramatically,

- Continental and polar ice sheets and alpine glaciers appear and/or expand,

- The air is drier, and

- Sea levels are lower as large volumes of liquid water become tied up in solid forms (i.e., ice, snow).

The Earth has experienced a number of Ice Ages; and it might surprise you to learn that it is technically still in the most recent one, which began about two million years ago.

Ice core analyses indicate that Ice Ages tend to develop more slowly than they end, which is due at least in part to the liberation of CO_2 from the ocean, thus reinforcing greenhouse effects. But while the weather can change within a few minutes, historical evidence indicates that even what are considered as "rapid" temperature changes over relative small regions take years or even decades.

Smaller-Scale Fluctuations

Within the major Ice Ages, however, are smaller-scale fluctuations that don't last nearly as long but can still produce quite significant temperature changes. These include colder "glacial periods" and warmer "interglacials." The Earth has been in an interglacial period for over 11,000 years, and such periods historically last for about 12,000 years; but some people maintain this one may last a bit longer due to an increased level of greenhouse gases. Stick around for another millennium or so, and find out!

* Pause briefly and review the above information on Natural Equilibrium, Temperature Change Triggers, Large-Scale Changes, and Smaller-Scale Fluctuations.

Global Warming Issues

The term "global warming" can be traced back to a 1975 *Science* paper by Wally Broecker of Columbia University entitled, *Climatic Change: Are we on the brink of a pronounced global warming?* but its meaning remains somewhat controversial. Taken literally, it seems to imply the average temperature of the entire planet is increasing, but it is considerably less clear whether it refers to surface or upper level temperature(s) of the air, the water, or both, and how supporting data are measured or derived. It is not the intent here to take sides, but rather to present information on issues others have raised to enable you to make better informed decisions about this subject, and we will begin by examining how one might go about:

Taking Earth's Temperature—If you plan to determine whether something is warming or cooling you first must be able to properly take its temperature, but how do you measure a *legitimate average* temperature of an entire planet? Let's examine several hypothetical examples to try to put this issue into perspective.

Suppose that you need to determine the *average* temperature of a flea to within tenths of a degree F. This may be possible with one measurement taken using a single (itsy-bitsy) calibrated thermometer placed anywhere upon the flea's body because of its tiny size. But suppose you need to determine the average body temperature of a seven-ton elephant. You may have some appropriately-sized and calibrated thermometers, but *how many* will you need and *where* do you need to place them to obtain a truly representative average temperature?

Now imagine that your subject is the entire Earth, and you start to appreciate the magnitude of the challenge. How do you confirm the measured temperatures are sufficiently accurate and valid? How many temperature sensors will you need, how should you distribute them about the globe to obtain truly representative data, and how will you adjust the measured data for local weather effects?

Determining an entire planet's average temperature requires more than a lot of calibrated thermometers. It also requires use of remote sensors (e.g., satellites), plus confidence in statistical methods and models used to quality control and to process the measured data. The following sections examine issues surrounding each of these factors more closely:

Data Accuracy—Because surface temperatures are frequently cited by both proponents and opponents of global warming, we will focus upon that data. It is generally accepted that "relatively reliable" surface temperature measurements have only been available since about 1850 over a relatively small portion of the planet, so earlier values and those in remote locations have to be inferred from other evidence like tree rings, fossils, etc., and such inferred values may be less precise than measured data. Since climate changes are typically expressed in tenths of a degree, such inaccuracies can be quite important.

Data Quality—But what does "relatively reliable" actually mean? The WMO provides calibration, sheltering, and placement standards for official surface air temperature measuring instruments, but recent studies reveal that a number of these instruments are: located near exhaust fans of air conditioning units, on hot rooftops, surrounded by parking lots, near sidewalks or buildings that can absorb and re-radiate heat, or now located in "urban heat islands." Such factors could artificially skew data measurements toward higher temperatures.

Data Distribution—Ideally surface air temperature measuring instruments would be evenly distributed about the globe, but that is far from the case. The Goddard Institute for Space Studies (GISS) is a primary provider of climate data, and has found nearly 70 percent of all reporting stations lie between 30 and 60 degrees North latitude and that half of those are located in the U.S.

This means that to compute a "global average" temperature one must extrapolate the available measurements to the rest of the planet, and no two groups currently do this exactly the same. It also means that values in data-sparse regions (e.g., the southern hemisphere) may be artificially influenced by extrapolations of data-rich areas (e.g., the northern hemisphere), and this can be problematic if actual temperatures in the regions are significantly different.

Data Density—Despite thousands of surface temperature reporting stations and the addition of satellite sensor-derived temperature data since 1979, the fact is that:

- A very substantial portion of the Earth is not adequately monitored, and

- Organizations handle data density problems differently.

For example, the data set used by the Intergovernmental Panel on Climate Change (IPCC) averages temperature anomalies in five degree square grid boxes; but in data-sparse regions, grid boxes of this size may contain very little, if any, actual data.

Data Adjustments—Before the raw data are averaged, various organizations apply different "adjustments" for: obviously bad data, varying observation times, equipment differences, reporting station histories, missing data, urban warming, etc. There are concerns that some of these adjustments artificially bias values toward higher temperatures in certain areas, and there is some evidence of this.

The National Oceanic and Atmospheric Administration (NOAA) has noted that the cumulative effect of all adjustments to the Global Historical Climate Network (GHCN) is roughly one-half of a degree F over the fifty-year period between the 1940s and the 1990s, which is similar to the amount of warming reported. At this point, it is also worthwhile to mention that while both the raw and adjusted data are publicly available from the U.S.-based GHCN, data used by the IPCC is not.

Data Averaging—Organizations also differ in the ways they calculate average temperature values. Some record the minimum and maximum temperatures at a given station for each day, and compute the daily averages; some use these to calculate monthly averages, and some compute annual averages from monthly ones. So what exactly does an average of an average of an average represent?

Organizations like NOAA's National Climatic Data Center (NCDC) do not actually use measured temperatures but rather *anomalies*, which are positive or negative departures from "long-term average reference values." This is done to try to filter out local effects and to compensate for a scarcity of temperature measurements in data-sparse areas (e.g., the Sahara Desert). However, some maintain that the validity of this method depends on the validity of the 'long-term reference values" chosen. If the chosen reference is *too warm*, one may expect positive anomalies to be smaller and the negative ones larger. Conversely, if one's chosen reference temperature is *too cool* the opposite might be true, which would lead to skewing the data toward warming. We'll let the mathematicians sort this one out, though.

Data Inconsistencies—There are three primary surface-measured temperature data sets, plus two satellite-derived temperature data sets that use identical sensor data, but different processing methods. A comparison of all five data sets between 1979 (when satellite derived temperature data first became available) and 1997 revealed that the satellite data indicated less warming. In fact, between 1979 and 1997 it indicated no significant warming; and after the El Niño period from 1997 to 1998 very little warming, except that induced by data adjustments.

Another often-cited inconsistency reinforces the proposition that warming may not be as "global" as some suggest. Although regions of the data-rich northern hemisphere appear to be warming, there have been reports of increased snow accumulations in Antarctica. And in their 2007 report, the IPCC also stated that, "Antarctic sea ice extent continues to show inter-annual variability and localized changes but no statistically significant average trends, consistent with the lack of warming reflected in atmospheric temperatures averaged across the region."

Data Selection—Do you remember our discussion about the importance of time frame selection when computing climatic values? Some have suggested that the IPCC might have skewed results by selecting a particularly cool year (i.e., 1976) as the starting point for calculating the predicted short-term linear warming trend.

Empirical Data—Some of the strongest global warming arguments are based on empirical (observational) data, which form a foundation for many well-respected hypotheses. Some glaciers and ice fields are shrinking and sea levels appear to be rising, but since the Earth has been in its current interglacial period for more than 11,000 years, isn't that what would be expected? There is nothing wrong with using empirical data, as long as one recognizes it is more qualitative than quantitative and is subject to the same limitations (e.g., observation quality, density, distribution) as measured data.

✳ **Take a break now to let these basic *Global Warming Issues* sink in.**

Even if all surface temperature data were perfect, it would still have to be correlated and compiled before it can be used as input to computer-based climate models, which conveniently brings us to our next subject of:

Climate Computer Models

Most are surprised to learn that after initialization, many models do not rely upon actual temperature data. Such models are basically mathematical interpretations of the Earth's climate system based on principles of: thermodynamics, radiative energy transfer, fluid dynamics, etc. While modelers may attempt to incorporate as many processes as possible, simplifications of the actual climate system are necessary because of available computer processing power and limitations in knowledge of the climate system.

Model Differences—Years in the environmental modeling community proved that model results can be "shaped" by: differences in their initializing data sets, different greenhouse gas inputs, varying model sensitivities, tuning adjustments, and other entering assumptions; and even the most honorable of modelers can become prejudiced toward models they develop or choose to use. This is not necessarily malicious, it is simply human nature.

Model Limitations—The IPCC considers a number of different climate models. Evaluations are done by comparing a model's ability to simulate past or current conditions, but none of the models simulates all aspects of the Earth's climate or accurately predicts all the effects of global warming.

The above statement is reinforced by the results of a recent study by a group of scientists that compared the warming predicted by six models used by the IPCC with actual satellite sensor data between 2000 and 2011. The study's details and findings that were published in a major peer-reviewed scientific journal revealed "a huge discrepancy" between the models' predictions and observed conditions. Measurements by the National Aeronautics and Space Administration (NASA) ERB and Terra satellites indicate that less heat is being trapped in the Earth's atmosphere and much more is escaping into space than these models predict.

Other Credibility Issues—The number of respected atmospheric scientists who are skeptical about IPCC credibility because of: the lack of term limits for IPCC leadership, apparent conflicts of interest, lack of transparency, and evidence of advocacy instead of truly unbiased analyses can't be ignored. These are rather important allegations because of the relationships between the IPCC and the U.S. Environmental Protection Agency (EPA).

✳ **Pause briefly and review information on *Climate Computer Models.***

The Greenhouse Effect

Few deny that this phenomenon exists, but there's no shortage of misinformation about it. For one thing, it is perfectly natural; and for another, it is not necessarily bad. Actually, without the greenhouse effect, the Earth would be cold, barren, and uninhabitable; and it has been functioning for as long as we have had an atmosphere.

Past Atmospheres. According to the Big Bang theory, the Universe began about 13.75 billion years ago, but there were not many decent clocks back then. Most modern scientists believe the Earth's atmosphere has changed dramatically at least three times before humans appeared. When the Earth was just a cloud of hot gases and dust, there actually was not very much of an atmosphere. As our spinning young planet cooled, its earliest atmosphere probably consisted of such gases as: Hydrogen, Helium, and Ammonia (NH_3); plus compounds of Chlorine, Sulfur, Fluorine, and Bromine. But this first atmosphere didn't last very long, and was stripped away by solar winds until a more steady state was reached.

Our planet's second atmosphere was mainly created by "out gassing" (think of Earth burping), and likely consisted of gases similar to ones emitted by today's volcanoes (e.g., Water Vapor, CO_2, Carbon Monoxide (CO), Sulfur Dioxide (SO_2), Sulfur, Chlorine, Nitrogen, Hydrogen, Ammonia, Methane), but no free Oxygen. As the Earth cooled off, the Water Vapor condensed to create the oceans, while photosynthesis decreased the level of CO_2 and produced atmospheric Oxygen (O_2), which in turn allowed the appearance of Oxygen-breathing life forms.

Our Current Atmosphere. Today, the thin envelope of gases and other stuff we call the Earth's atmosphere consists of about: 78 percent Nitrogen, 21 percent Oxygen, 0.9 percent Argon, 0.04 percent CO_2, plus smaller percentages of other gases, including Water Vapor. It is interesting to note that the atmospheres of Mars and Venus are both dominated by CO_2, and both those planets exhibit very limited industrial activity.

Earth receives solar energy in ultraviolet, visible, and near-infrared wavelengths. Roughly half of it gets absorbed at the surface and is re-radiated back upward as longer (far-infrared) wavelengths. Some atmospheric gases absorb and re-radiate this long-wave infrared energy in all directions, and the amount that is radiated back downward produces the so-called "greenhouse effect."

Greenhouse Gases. The principal "greenhouse gases" include: Water Vapor, CO_2, Methane, and Ozone (O_3). When ranked by their direct contribution to the greenhouse effect, Water Vapor contributes about 36 to 72 percent, CO_2 9 to 26 percent, Methane 4 to 9 percent, and O_3 3 to 7 percent. However, the overall Global Warming Potential (GWP) of any gas is determined by its: direct effectiveness, abundance, atmospheric lifespan, and indirect effect(s).

At the molecular level, the direct effect of Methane is about seventy-two times that of CO_2, and it has a large indirect effect because it contributes to formation of O_3; but Methane is less abundant and has a shorter atmospheric lifespan. It may not be as abundant as Water vapor or have the direct effects of Methane, but most of the focus these days is upon CO_2, because of its relatively long atmospheric lifespan (or perhaps because studying Methane literally stinks).

Atmospheric Particulates. We also need to discuss atmospheric particulates, many of which are produced by combustion of fossil fuels, and can cause either heating or cooling effects. Some claim the increase in CO_2 and "dimming" of the Sun's energy by atmospheric particulates produced by volcanic activity, wildfires, and industrial pollution, have basically offset each other since 1960.

By this time, I realize all of you are incredibly anxious to learn about *soot*, which can cool or warm, depending upon whether it is airborne or deposited. Airborne soot directly absorbs solar radiation, which heats up the atmosphere and cools the underlying surface. In locations that produce high amounts of soot, like rural India, up to 50 percent of the surface warming caused by greenhouse effects may be offset by atmospheric soot. When it is deposited upon glaciers or ice fields in polar regions, however, that same soot can cause surface warming.

Politics and Ideologies

A Gallup survey of 127 countries in 2007 and 2008 revealed that over one-third of the world's population was unaware of global warming. In 2009, the Pew Research Center conducted a survey which indicated the perception that global warming was a serious problem had decreased among the general public in the U.S.; and a 2011 Rasmussen poll revealed 69 percent of adults in the U.S. believed it is at least somewhat likely some scientists have falsified global warming research.

One reason might be that global warming has definitely become a political issue, and public opinion of politics is at a historic low. The Kyoto Protocol is essentially an international treaty proposed by the United Nations Framework Convention on Climate Change (UNFCCC) to limit greenhouse emissions. It was not ratified by the U.S., because it did not apply firmly enough to large developing nations (and major greenhouse gas emitters) including China and India; and in 2011, Canada announced that it would also withdraw from the Kyoto Protocol.

The politics of global warming has included intense lobbying efforts and funding from special interest groups from all sides, which can influence legislation and policy decisions. A 2007 study showed that roughly 95 percent of Congressional Democrats surveyed felt that, "it has been proven beyond a reasonable doubt the Earth is warming because of man-made problems," while only 13 percent of the Republicans surveyed agreed. In 2011, another survey indicated that varying majorities of Democrats, Independents, and Republicans felt there was evidence of warming (for some reason), but this is as close to a consensus as it gets.

Ideological biases have also crept into this issue, further straining the credibility of extremists on all sides. This might be common (and even accepted) in some circles, but evidence of it among some elements of the scientific community is particularly concerning and totally unacceptable.

* **Take a break to review the above information about** *The Greenhouse Effect, Politics & Ideologies.*

History Lessons

History can teach us valuable lessons if we are willing to learn them, and there is historical evidence that the following events are not as rare as recent publicity would lead us to believe:

Temperature Swings

According to NASA's Goddard Institute for Space Studies (GISS), during the past billion years the Earth experienced glacial periods 925, 800, 680, 450, 330, and 2 million years ago. During the latter *Pleistocene Ice Age*, ice sheets covered much of Europe, Asia, and North America for extended periods of time; and the global temperatures were about seven to nine degrees F colder than they are today. But as previously mentioned, over 11,000 years ago we entered a warmer interglacial and these ice sheets began to retreat.

The current interglacial was briefly interrupted by a 1,500 year-long cooler period, but resumed until between about 5000 and 3000BC when temperatures were about two to four degrees F warmer than they are today. It is also interesting to note that during this so-called "Climatic Optimum" period, a number of the Earth's great ancient civilizations began to flourish.

During another cooling period between about 3000 and 2000BC, lower sea levels caused the emergence of some islands (e.g., the Bahamas) that remain above water. Another short warming period from around 2000 to 1500BC preceded yet another cooling period between 1500 and 750BC when the renewed growth of glaciers caused sea level to drop about six to nine feet below the current level.

There was another warming period from roughly 750 to 150BC, followed by a cooling period lasting until around 900AD and during which the Nile froze. The "Little Climatic Optimum" between 900 and 1200AD was actually so warm that the snow level in the Rockies was over 1,200 feet higher than it is today. This was followed by a period of cooler temperatures and extreme weather events that included: floods, large seasonal temperature variations, and a drought in the American Southwest between 1276 and 1299AD. But wait, there's more!

Cold temperatures between 1550 and 1850AD led scientists to call that period "The Little Ice Age." They were so low that between 1753 and 1759AD roughly 25 percent of Iceland's population died from crop failure and the newspapers in New England called 1816 "the year without a summer"; but since approximately 1850AD, we have been experiencing another warming period. If we are simply willing to learn, history can teach us that large-scale temperature fluctuations actually are not so unusual, and we probably have more of them ahead.

Tipping Points

There is a lot of talk these days about positive forcing and feedback mechanisms and the chance of greenhouse gas concentration reaching a "tipping point" when warming would be further accelerated. Although this is possible, it is important to note that completely natural events have caused the Earth's atmosphere to reach and exceed such tipping points throughout history, and, in all those cases, nature has reversed course to shift back toward a state of equilibrium.

Mass Extinctions

Mass extinctions tend to be unpopular, particularly among species that become extinct; but they are not as rare as many folks think. There is scientific evidence there have been at least five great extinctions in the last half billion years, plus other smaller events. About 500 million years ago, 1/2 of all the current animal families became extinct; while a mass extinction roughly 345 million years ago wiped out another 30 percent.

Around 250 million years ago, 50 percent of all the current animals, 95 percent of the marine species, and many amphibians and trees became extinct. Roughly 180 million years ago, 35 percent of the current animal families, including many reptiles and marine mollusks were wiped out, and 65 million years ago most dinosaurs and more mollusks became extinct. All of these were due to totally natural causes, but some subsequent extinctions have been influenced by human activities.

Humans have destroyed and polluted habitats, introduced non-native species and diseases, and overexploited plant and animal populations; but extinctions might not be entirely bad. For example, great populations of large herbivores produce significant amounts of the greenhouse gas Methane, and evidence indicates that both natural events and human activities (e.g., hunting) which reduced such populations also reduced the level of atmospheric Methane.

* **Pause to review what you've learned by these _History Lessons_.**

What We Know

The main reason hindsight is typically better than foresight is that we know what happened if we simply pay attention, so let's summarize what we actually know at this point about climate change and global warming.

- From their official WMO definitions, we know that _weather_ describes actual atmospheric conditions, while _climate_ represents long-term statistical values.

- From these same official definitions, we also know that both weather and climate are typically viewed as local or regional, not global, characteristics.

- We know that both the Earth's atmosphere and temperatures have undergone major changes due to natural causes and such changes are likely to continue.

- We also know that future changes will probably be more hospitable to some life forms and less hospitable to others.

- We know that our planet is still technically in its last Ice Age, albeit in a warmer interglacial period which began more than 11,000 years ago.

- We also know that at least empirically the surface temperatures in some but not all areas appear to be increasing.

- We know that computing a representative average temperature of an entire planet is very complex and the practical value of such a number has been debated.

- We also know that the quality, density, distribution, adjustments, selection, and inconsistencies of some temperature data have been questioned.

- We know that current global climate models are incomplete and sensitive to initial data set validity, tuning differences, and other entering assumptions.

- We also know that the credibility of models that have been subjected to very thorough and objective peer reviews and initialized with valid data sets, may be less questionable than the objectivity of some individuals who promote their use.

- We know that the greenhouse effect is one of at least six principal factors which can and have triggered major temperature changes throughout history.

- We also know that human activities and natural events can have both positive and negative effects upon temperature.

- We know that public opinion, politics, and ideological biases all play roles in the global warming controversy and may have affected scientific positions.

- We also know claims that there is currently unanimous or general acceptance of all aspects of global warming throughout the world's scientific community are simply not true.

We know that the Earth's natural environment will continue to change. We do not yet know how significant any effect(s) of human activities might have upon the climate, or what the eventual outcome of this latest episode in Mother Nature's never-ending balancing act will be.

The Bottom Line

Greed, corruption, and agenda-biased pseudo-science have played major roles in keeping this dispute alive. As a former oceanographer and meteorologist, I can appreciate healthy scientific debate, but extremists on all sides have very serious credibility issues. Experience should teach us that to properly resolve this matter we need to replace the politics and ideologies with better leadership and more common sense. At this point you know much more about the weather, climate, temperature change, and global warming than most people, so go baffle them with your brilliance.

✱ **Take a break and review this Chapter about *Climate Change*.**

CHAPTER THREE:
CONVERSATION STARTERS

Exercise Guide: This "cool down" Chapter provides relief from mental stresses created in Chapter Two. It is fairly easy and entertaining, but you still might want to pause briefly at the points indicated. And even though the subject may at first seem silly, communicating effectively is a very essential modern survival skill.

If you still are not convinced our communication skills need exercising, just try carrying on coherent conversations with people plugged into their Ear buds or Blueteeth, who can't spell, type with their thumbs, and use "chat-speak" terms like OMG and LOL. If our power grid suffered a major failure, those folks will probably wander about as itinerant bands blankly staring at the equally blank screens of their iPods or Smart Phones. Like IMHO, that's awesome, Dude!

Even though communicating is one of the most important things we all must do, many people have a little trouble getting started. Opening lines like, "Have you always had that drooling problem?" or "How long have you been out of re-hab?" may not always be appropriate and can sometimes kindle slightly more physical responses than gentle repartee. The exercises in this Chapter will help you hone your communications skills, while initiating your conversations with the following witty and provocative questions is certain to make others:

- Irresistibly attracted to your mental prowess, or

- Think you recently escaped from a nearby nut hut.

To avoid the latter, it might be prudent to reserve some of the following lines for discussions that have been thoroughly lubricated with adult beverages. Most of them are appended by sufficient information to make you sound like you actually know something about the topic, and they are cleverly categorized and listed in alphabetical order for your rapid reference in those unanticipated conversational emergencies.

Accidents

It may not be polite to dwell on others' misfortunes, but it can certainly stimulate some intriguing discussions. The following opening lines reflect actual statistics, and include handy safety tips.

> **Aircraft**—*Are you aware that most airplane crashes occur during takeoffs or landings?* Safety Tip—avoid traveling on flights that involve such maneuvers.

> **Automobiles**—*Do you realize that most automobile accidents occur within three miles of one's home?* Safety Tip—always park at least three miles from home.

> **Ballpoints**—*Did you know every year an average of a hundred people choke to death on ballpoint pens?* Safety Tip—ballpoint pens are not edible.

> **Crosswalks**—*Are you aware that statistics indicate an inordinate number of the pedestrians struck by vehicles*

are in marked crosswalks? Safety Tip—marked crosswalks are apparently dangerous, so avoid them if possible.

Leaners—*Do you realize that at least 250 people have fallen off the Leaning Tower of Pisa?* Safety Tip—don't lean over so far next time.

Night People—*Did you know that if you work nights, you are nearly twice as likely to have an accident as a day worker?* Safety Tip—get to work before sundown, and don't leave work until after sunrise.

Southpaws—*Are you aware that more than 2,500 left-handers are killed each year while trying to use products that are designed for right-handed people?* Safety Tip—make all products only for left-handers until things even up a bit.

Volcanoes—*Do you realize that at least 300,000 people have been killed by volcanoes over the last 500 years?* Safety Tip—keep away from places with molten rocks and poisonous gases.

Animals

These questions are excellent for kicking off conversations, because everyone thinks they know something about animals. Here are a few things, however, that they might not already know:

Armadillos—*Are you aware that armadillos can get malaria?* They can also carry leprosy, and such traits can affect their popularity as household pets.

Bats—*Did you realize that plants which rely heavily on bats to pollinate their flowers or spread their seeds* *include: bananas, cashews, dates, mangoes, figs, and the blue agave used to produce tequila?* Holy Bat Margaritas!

Camels—*Do you know that the Army created the U.S. Camel Corps in 1855 because of the camel's endurance, carrying capacity, and theory they would terrify Indians' war ponies?* By 1866, it was disbanded due to lack of interest and the fact that the camels were scaring the Cavalry's own pack animals.

Crocodiles—*Are you aware that 99 percent of baby crocodiles get eaten in their first year by large fish, herons, lizards, or adult crocodiles?* Thank goodness for large fish, herons, lizards, and adult crocodiles.

Giraffes—*Did you realize that giraffes have the highest blood pressures of all mammals?* Their twenty-five pound hearts produce blood pressures that are twice that of other mammals.

Hippos—*Do you know bull hippos often fend off rivals by flinging feces as far as they can with their fan-shaped tails?* The feces-flingers return to their hippo harems, while the splattered suitors head to the river for badly needed baths.

Hummingbirds—*Are you aware that hummingbirds cannot walk or hop?* But they can scoot sideways when they are perched.

Lions—*Did you realize that female lions are the principal hunters, while male lions are primarily concerned with defending their pride?* This seems rather similar to human behavior.

Lobsters—*Do you know that lobsters were once so plentiful that they were considered "poverty food," and*

Native Americans used them as fertilizer? In the 1800s lobsters were harvested by "Smack men," who sailed small boats called Well Smacks.

Mice—*Are you aware that male mice utter ultrasonic squeals of excitement when they encounter the scent of a receptive female?* Female mice don't do this, but they do chirp when meeting their female chums.

Monkeys—*Did you realize that male monkeys go bald just like male humans?* A parallel study revealed that male humans go bald just like male monkeys.

Mosquitoes—Do you know that *disease-carrying mosquitoes are responsible for more human deaths worldwide than any other animal?* The World Health Organization (WHO) estimates they cause over two million deaths a year.

Oysters—Are you aware *that many oysters are males during their first year of life and become females by age two or three?* There is fossil evidence that the Ancient Romans cultivated oysters more than 2,000 years ago, but little evidence that the Ancient Romans changed their sex as much as oysters.

Pelicans—*Did you realize* that *pelicans stand on their eggs to incubate them?* Eggs, with shells weakened by the effects of DDT, were often crushed by their own parents.

Polar Bears—*Do you know that Polar Bears have black skin and transparent guard hairs in their outer layer of fur?* They are also so well-insulated they can become overheated at temperatures above fifty degrees F.

Seahorses—*Are you aware that seahorses are also rather confused about their roles in the reproductive process?* Females deposit their eggs into the male's brooding pouch where they are fertilized, and in roughly eight to ten days the male goes into labor and squirts out little sea colts one at a time.

Sloths—*Did you realize that it can take a sloth up to a month to digest a meal?* Because of this, at any given moment roughly two-thirds of a sloth's body weight is due to its stomach contents.

Termites—*Do you know that some termites communicate by banging their heads against walls of their tunnels?* Some humans exhibit similar behavior.

Turtles—*Are you aware that to fit their heads, legs, and tails completely inside their shells, turtles have to exhale a little bit.* Some of them can also breathe through their butts, but that's another story.

Whales—*Did you realize that sperm whales are deep divers, reaching depths of over several thousand feet and staying submerged for more than an hour?* They also periodically expel stinky blobs of ambergris, a valuable ingredient in some French perfumes that is also sold as an aphrodisiac in the Middle East.

Arty Facts

Art objects may not be as cuddly as puppies, but they are easier to housebreak. The next time you need to strike up a conversation with an art lover huddled in a corner of a theater or gallery, try some of the following questions:

Movies—*Did you know that:*

- *The producers of "Gone With The Wind" were fined $5,000 for allowing the word "damn" to be spoken in the film?* And after Gary Cooper turned down the role of Rhett Butler he said, "Gone With The Wind is going to be the biggest flop in Hollywood history!"

- *The movie "Grease" was released in Venezuela under the title, "Vaselina"?* And in Spain it was released as "Brillantina" because its English title translated into "Grasa," or "Fat" in Spanish.

- *The mechanical shark in the movie "Jaws" was called Bruce, after Steven Spielberg's attorney?* There were actually three Bruces, one of which was nicknamed "The Big &%#," which rhymes with "The Big Curd."

- *Alfred Hitchcock used chocolate syrup to portray blood during the shower scene in his famous black-and-white film "Psycho"?* And after she saw the movie, Janet Leigh reportedly decided to take baths instead of showers.

Paintings—*Did you realize that:*

- *There were no portraits painted of Christopher Columbus during his lifetime?* All the paintings of Chris were done from written descriptions of his physical appearance.

- *The Mona Lisa has no eyebrows because during the Italian Renaissance it was fashionable to shave them off?* And because the famous painting is considered priceless, it's also uninsured.

- *Pablo Picasso was baptized Pablo Diego José Francisco de Paula Juan Nepomuceno María de los Remedios Cipriano de la Santísima Trinidad Martyr Patricio Clito Ruíz y Picasso?* No wonder we just call him Pablo!

Sculpture—*Are you aware that:*

- *Roman statues were often constructed with heads that could be removed and replaced?* Marble wasn't cheap back then, and this made it easier to remain in step with changes in political regimes and/or deities.

- *The golden statues atop most Mormon temples depict the Angel Moroni?* This is the angel who told Prophet Joseph Smith where to find the buried records of an Ancient civilization that became the Book of Mormon.

- *Sculptor Michelangelo was considered rude, short-tempered, and difficult to work with?* He also didn't like Leonardo d Vinci much at all.

✳ **Pause to practice your lines about *Accidents*, *Animals*, and *Art*.**

Beverages

There are all sorts of opportunities to use the opening lines under this category. You might be at a luau or a house party, or simply longing to strike up another intellectual discussion about the physics of fruit in carbonated beverages at your neighborhood pub. In any case, here are a few guaranteed conversation starters.

Bobbing Raisins—*Do you realize that a raisin dropped into a glass of fresh Champagne will bob up and down?* This is because the carbonation bubbles get trapped in the raisin's wrinkled skin, then pop when it reaches the surface.

Glowing Ice Cubes—*Do you know how to make glowing ice cubes?* Just fill your ice trays with a fifty-fifty mixture of tonic water and regular water to make cubes that have a bluish glow under ultraviolet and some fluorescent lights. If anybody asks, just tell them that you get bottled water from Three Mile Island.

Mai Tai History—*Are you aware of where and how the Mai Tai originated?* Although there is some controversy about this, the most popular version says it was created in 1944 by "Trader Vic" Bergeron in his Oakland, CA. When he offered it to some friends who were visiting from Tahiti, they exclaimed "Maita'i roa ae!" which as you all know, means "very good!"

Celebrities

If you have the nerve or the lack of good sense to strike up a conservation with a tabloid journalist or gossip columnist, the following opening lines might help:

Johnny Carson—*Did you know that the famous talk show host's first paid gigs were at age fourteen, when "The Great Carsoni" performed magic acts for the local Rotary Club meeting, Methodist church socials, and his mother's bridge club?* And whenever Johnny's opening monologue bombed, the Late Show band would usually start playing "Tea For Two."

Joan Crawford—*Are you aware that each of the three times this famous actress remarried, she changed the name of her Brentwood estate?* And at such times, Mommie Dearest also replaced all of its toilet seats.

Howdy Doody—*Do you realize that Howdy Doody had a twin brother named Double Doody?* He also had a sister named Heidy Doody.

Albert Einstein—*Did you know that the phamous physicist's second wife, Elsa, was also his maternal first cousin?* And to further keep things in the family, she was also his paternal second cousin.

Superman—*Did you know that in the comic books, Superman had a monkey name Beppo and a horse named Comet?* And his real parents were Jor-El and Lara, who perished when the planet Krypton was destroyed.

The Lone Ranger—*Are you aware the Lone Ranger's mask was supposedly made out of the vest of his dead brother, Captain Dan Reid, by Tonto?* And after retiring from playing Tonto, Jay Silverheels became a harness racing driver.

The Marx Brothers—*Do you realize that the stage names of the five Marx brothers were Groucho, Chico, Harpo, Gummo, and Zeppo?* But their real names were Julius, Leonard, Adolph, Milton, and Herbert, respectively.

Chemistry

When it comes to being wild and crazy, some insist chemists are outdone only by astrophysicists and performance artists. So the next time you're invited to a party in the laboratory, here are a few conversation starters that should come in handy:

It's in the Name—*Do you know what the name WD-40 stands for?* It was the result of the fortieth try by the Rocket Chemical Company to create a Water Displacement (WD) solvent, and was first used to protect missile skins from corrosion.

Liquid Metals—*Do you know if there are any metals that are liquids at room temperature (i.e., seventy-seven degrees F)?* Mercury and Bromine are both liquid, while Francium, Cesium, and Gallium will also melt by the time the temperature reaches eighty-seven degrees F.

Stumbling Into Putty—*Do you realize what the chemist who discovered Silly Putty was actually trying to create?* It was synthetic rubber, because in 1943 natural rubber was being rationed due to World War II.

Viagra Uses—*Are you aware that Viagra can keep flowers from wilting and in bloom for up to a week longer?* It reportedly has other uses as well.

Crime

Striking up conversations with gang members, drug kingpins, or career politicians is not easy unless you also know some things about organized crime. Here are a few ways to begin:

Al Capone—*Did you know that Al's most intimate friends called him "Snorky"?* It was slang for "elegant," and nobody called him Scarface more than once.

Bulletproof Vests—*Are you aware that the bulletproof vest was invented by a pizza delivery guy?* In 1969,

Richard Davis invented it to protect himself during deliveries in Detroit.

Dillinger's Cars—*Did you realize that John Dillinger only used Fords for his getaway vehicles?* Both he and Clyde Barrow are reported to have written letters to Ford praising their vehicles, but when Dillinger was ambushed by federal agents outside Chicago's *Biograph Theater* in 1934, he was on foot.

Electric Chair—*Do you know that the idea for the electric chair came from a dentist?* It came to Dr. Al Southwick, DDS back in 1881 after he saw a drunk stumble onto the live terminals of a power generator in Buffalo, NY.

Gunfighters—*Are you aware that more cowboys died while crossing rivers than in gunfights?* Safety Tip— pick gunfights on your current side of a river.

Mona's Missing!—*Did you realize that when the Mona Lisa was stolen from the Louvre in 1911 some thought the theft was financed by a wealthy collector, others claimed it was a German plot to discredit France, and a disgruntled poet even tried to implicate Pablo Picasso?* It turned up two years later in Florence, where a former Louvre employee confessed to trying to return Mona to her country of origin.

Prison Sentences—*Did you know that the longest prison sentence given to date in the U.S. is 10,000 years?* An Alabama court gave it to Dudley Wayne Kyzer in 1981 for murdering his wife, mother-in-law, and a college student.

✳ **Pause and review the information in the last several sections.**

Food

The opening lines in this next category are extremely popular because most of us regularly ingest some sort of sustenance. Be warned, however, that the following questions may cause unexpected reactions in supermodels and other bulimics:

Animal Crackers—*Are you aware that there currently are nineteen different shapes in Barnum's Animal Crackers zoo?* And the strings on those boxes were originally there to hang them on Christmas trees.

Bananas—*Did you realize that foil-wrapped bananas were first sold (for ten cents each) to the American public at Philadelphia's Centennial Celebration in 1876?* Before then, sailors returning from the Caribbean smuggled them in, because carrying bananas aboard ship was considered to be bad luck.

Cereal—*Do you know that the average American eats roughly twelve pounds of cereal per year?* Its name comes from Ceres, the Roman goddess of grain.

Graham Crackers—*Are you aware that these were invented by Presbyterian Minister Sylvester Graham in 1829 as a health food to suppress carnal urges and were initially marketed as "Dr. Graham's Honey Biskets"?* The Reverend Graham also helped found the American Vegetarian Society, and lived to the ripe old age of fifty-seven.

Happy Meals—*Do you realize that 40 percent of McDonald's profits come from the sales of Happy Meals?* The first Happy Meals back in 1979 included a hamburger, fries, small drink, cookies, and a toy for just one dollar.

Jell-O—*Did you know that more lime-flavored Jell-O is consumed in Salt Lake City than in any other city in the U.S.?* And the original flavors back in 1895 were strawberry, orange, raspberry, and lemon.

Ketchup—*Are you aware that in the 1800s, people used ketchup but wouldn't eat raw tomatoes because they believed the latter were poisonous?* And according to Heinz, only about 11 percent of us know about tapping the "57" sweet spot on the neck of their iconic glass bottle to help the ketchup pour.

M&Ms—*Do you realize that when M&Ms were introduced in 1941, they were sold only to the military?* They were originally packaged in cardboard tubes.

Olives—*Did you know that in 1987 American Airlines saved over $40,000 by eliminating an olive from every salad served in first class?* While olive trees can live for millennia, their average age is only about 500 years.

Peanuts—*Are you aware that roughly 99 percent of all peanuts grown in the U.S. today come from seven states (Georgia, Texas, Alabama, Oklahoma, Florida, Virginia, and North Carolina)?* Other names for peanuts are: goobers, earth nuts, monkey nuts, and ground nuts; but they are closely related to peas.

Pickles—*Do you realize that Julius Caesar fed his men pickles to give them strength?* And Cleopatra attributed her beauty to eating copious cucumbers.

Potato Chips—*Did you know that potato chips were invented in 1853 after a chef got angry at a diner who kept sending his French fries back because they were too soggy, thick, and bland?* Today they are the most popular snack food in the U.S.

Raisins—*Are you aware that the commercial potential of raisins in California was discovered by accident when a San Francisco grocer marketed grapes shriveled by a hot spell in 1873 as, "Peruvian Delicacies."* Today, California grows half of all raisins sold, and the nickname of Fresno's first professional baseball team was "The Raisin Eaters."

Snickers—*Did you realize that the Snickers candy bar was named after the Mars family's favorite horse?* It's also the most popular candy bar in America.

Sundaes—*Are you aware that because religious leaders objected to drinking soda water on Sundays, an Illinois druggist left it out of his confections, and the ice cream sundae was born?* Today most ice cream is eaten on Sundays, and between nine p.m. and eleven p.m.

Gambling

You never know when you may need to start up a conversation with *Bernie the Bookie*, or a group of grannies on a weekend junket to Reno. Even if you don't know when to hold 'em and when to fold 'em, the following openings can help:

Casino Lingo—*Did you know casino employees call high-stakes gamblers "Whales" and inexperienced ones "Fish"?* And consistent with this nautical theme, they also call the gamblers who arrive in buses "*Boat People.*"

Playing Cards—*Are you aware that the King of Hearts is the only King without a moustache in a deck of standard playing cards?* And do you also realize the King of Spades actually represents biblical King David, the King of Diamonds depicts Julius Caesar, the King of Clubs represents Alexander the Great, and the King of Hearts depicts Charles the Great (also known as Charlemagne)?

Slot Machines—*Did you know that in Las Vegas there is one slot machine for every eight residents?* Car mechanic Charles Fey invented the first one called the Liberty Bell in 1895, and for years they were called Bell machines because a combination of three Bells paid the grand jackpot of fifty cents (in nickels).

Geography

I realize geography may not be a topic that immediately comes to mind unless you are in a center seat on an airplane headed to a Cartographer's Convention, or on a road trip in a minivan full of antsy adolescents whose idea of intellectual dialogue is another round of "Ninety-Nine Bottles of Beer on the Wall." In such cases, desperation frequently overcomes sanity and the following seem a lot more appropriate:

Continental Quiz—*Did you realize that the names of all seven continents begin and end with the same letters?* And in alphabetical order they include: Africa, Antarctica, Asia, Australia, Europe, North America, and South America.

Road Room—*Are you aware that every mile of a four-lane freeway takes up over seventeen acres of land?* And in case you are still curious, there are almost four million miles of roads and streets in the U.S.

Rods & Furlongs—*Did you know that in medieval times they surveyed land using standard metal rods 16.5 feet long, and a furlong was the 40-rod length of a furrow in a communal field?* These days, rods are used to indicate portage distances on canoeing maps because they are roughly the same length as a typical canoe, while furlongs are still used for describing distances in horse racing.

The Seven Seas—*Can you name the world's seven seas?* They have changed a little over the years; but according to NOAA, the current ones are the: Arctic, North Atlantic, South Atlantic, North Pacific, South Pacific, Southern, and Indian Oceans.

What's An Acre?—*Do you know how large an acre really is?* Acres don't have a standard shape, but an acre contains an area equal to that in a rectangle that is 4 rods (66 feet) wide and 40 rods (660 feet) long. It's roughly the amount of land an average ox could plow in a day, in case you own an average ox.

✳ **Pause to practice these opening lines and rest your ox.**

Inventors

Speed dating at the U.S. *Patent Office's Roundhouse Cafe* is a good way to meet eligible inventors. The next time you try it, use the following conversation starters to impress them:

Alexander Graham Bell—*Are you aware that the inventor of the telephone would not allow one in his study because its ringing broke his concentration?* And the famous first telephone sentence, "Mr. Watson come here, I want you!" was actually spoken because Bell had spilled acid onto his pants.

George Eastman—*Did you know that the founder of Kodak didn't like having his photograph taken?* He also was an avid bicyclist who thought brakes were dangerous, so he dragged his feet on the ground when he wanted to stop.

Thomas Edison—*Do you realize that the patent holder for the light bulb was considered to be "a dull student" and even "addled" by his teachers?* Edison proposed to his second wife in Morse code, and some of his lesser-known inventions include the electric tattoo machine and concrete furniture.

Names

As all of us learned in pre-school, *onomastics* is the study of names. Sometimes circumstances cry out for conversations about it, in which case the following will prove invaluable.

The Last Animal—*Are you aware that the last animal listed in an English dictionary is the zyzzyva?* It's a tropical weevil, but you already knew that.

The Longest Place—*Did you know that the longest official place name is Taumatawhakatangi hangakoauauotamateapokaiwhenuakitanatahu?* It's a hill in New Zealand.

Punctuation

Sooner or later we all must punctuate, and if we live long enough we may even get a chance to use a semicolon correctly. Try the following questions the next time you want to strike up a conversation with somebody who realizes there is more to punctuation than making a :)

Do You Interrobang?—*Do you realize that a new punctuation mark called the interrobang was created in the 1960s?* It is a combination of an exclamation and a question mark and is used to end a question that is asked excitedly.

Exclamation Marks—*Are you aware that typewriters didn't have any keys for exclamation marks until the 1970s?* You created the latter by first typing a period, then backspacing and typing an apostrophe over it.

What's It Called?—*Did you know that the dot atop the lower-case letter "i" is called a "tittle?"* Of course you did.

Sports

It can be tough to have a conversation in a Sports Bar, when there are twenty-six large-screen TVs all blaring with various games, matches, tournaments, playoffs, and color commentators rehashing what happened. But the next time a commercial comes on, try one of the following ice-breakers:

Baseball—*Are you aware that the average life span of a major league baseball is only five to seven pitches?* And the longest official throw was by a Canadian minor leaguer, who heaved a ball 445 feet and 10 inches before it hit ground.

Football—*Did you know that football has more official rules than any other major American sport?* And the first intercollegiate game was played in 1869 between Rutgers and Princeton, with twenty-five players on each side.

Golf—*Do you realize that more people die while paying golf than any other major sport?* The leading causes of such deaths are heart attacks and strokes.

Ping Pong—*Are you aware that during a ping pong tournament in 1936, the opponents volleyed for more than two hours on an opening serve?* Ping Pong began as an English parlor game played upon dining room tables and was called by various names including: Gossima, Flim-Flam, and Whiff-Whaff.

Word Stuff

Words are excellent choices for starting sentences, which is why this section is provided. For more impact, you might wish to accompany them with irrelevant gestures.

About 40—*Did you know that the word forty is the only number in the English language in which all of its letters are in alphabetical order?*

Backwords—*Do you realize that the letters in "I'm a lasagna hog" written backwards spell "go hang a salami?"*

Now Hear This—*Are you aware that the word "listen" contains the same letters as the word "silent"?*

Rhyme These—*Did you know there are no words in the English language that rhyme with month, orange, purple, or silver?*

Shifting Hands—*Do you realize the longest English word that alternates hands when typed on a QWERTY keyboard is "skepticisms"?*

Triple Doubles—*Are you aware that the only word in the English language with three consecutive double letters is "bookkeeper"?*

Unrepeatable—*Did you know the only fifteen-letter English word that can be spelled without repeating a letter is "uncopyrightable"?*

World Leaders

Biographers tend to know a lot of stuff about famous folks and once primed, can rattle on *ad nauseum* about subjects with whom they are familiar. But if you are fearless (or foolhardy) enough to strike up a conversation with one of them, the following may be useful. Good luck!

Winston Churchill—*Did you know that British Prime Minister Churchill was born prematurely when his mother was attending a party at Blenheim Palace?* And he died at age ninety on the very same day of the year his father had passed away seventy years earlier.

Grover Cleveland—*Do you realize that U.S. President Cleveland's nicknames were "Big Steve" and "Uncle Jumbo"?* He married his god-daughter and one of his famous quotes is: "Sensible and responsible women do not want to vote."

Dwight Eisenhower—*Are you aware that U.S. President Eisenhower had pajamas with five stars on the lapels?* Perhaps it was because he was the Supreme Commander of Allied Forces in Europe during World War II.

King George III—*Did you know that on July 4, 1776, King George III of England wrote in his diary, "Nothing of importance happened today."* He apparently missed adoption of the U.S. Declaration of Independence.

Benjamin Harrison—*Did you realize U.S. President Harrison was afraid to touch the new-fangled White*

House light switches? But not for long, because he contracted pneumonia a month after his inauguration and died that April.

Thomas Jefferson—*Are you aware that Jefferson was the first U.S. President to shake hands in greeting others?* Washington and Adams both bowed.

Abraham Lincoln—*Do you realize that Lincoln was the first U.S. President to be born outside the original thirteen states?* Honest Abe was born in Kentucky.

Richard Nixon—*Did you know that Richard Millhouse Nixon was the first U.S. President whose name contained all of the letters in the word "criminal?"* The second was William Jefferson Clinton.

Ronald Reagan—*Are you aware that Reagan was the only U.S. President to head a labor union?* In 1959-60 he was president of the Screen Actors Guild.

Franklin D. Roosevelt—*Did you realize that FDR used Al Capone's car?* He first did so to give his "Day of Infamy" speech to Congress after the attack on Pearl Harbor, since Al's confiscated car was the only bulletproof one available.

Theodore Roosevelt—*Did you know that U.S. President Roosevelt drank about a gallon of coffee a day?* He also coined the phrase "good to the last drop," which became a slogan for Maxwell House coffee.

Harry S Truman—*Are you aware that U.S. President Truman didn't have an actual middle name?* This is why there is no period *after the "S" in his name.*

Martin Van Buren—*Did you know that Van Buren was the first U.S. President to have never been a British subject?* Martin's father ran a tavern in New York.

George Washington—*Did you realize Washington owned foxhounds named Drunkard, Tipsy, and Tipler?* And his second inaugural address was just 133 words long and only took him 90 seconds to deliver. Those were the days!

Eschewing Obfuscation

The thinly veiled message behind this Chapter is that communicating effectively isn't easy and requires practice. Modern society has lowered the bar with regard to acceptable standards for grammar and spelling, and conversational courtesy could also use work. How do you feel when someone interrupts talking with you to reply to his or her Dingleberry? Do you enjoy having to listen to half of a cell phone conversation in a restaurant, and how comfortable are you when you notice the driver of the big truck tailgating you is babbling on his Bluetooth?

Please help ensure that future species of humans don't have prehensile ears, large thumbs, and vestigial brains. If you never use any of these conversation starters, you will miss seeing some priceless expressions on the faces of your listeners. And if you happen to find yourself seated between an archaeologist and an anthropologist at your next dinner party, the next Chapter will help you come up with something incredibly stimulating to say even then.

✸ **Take a break to go impress folks with your conversation skills.**

CHAPTER FOUR:
FOOTPRINTS AND ECHOES

We've had three dogs that all loved to go for rides. One would intently stare out through the windshield like he wanted to drive, another preferred to sit backward and look out through the rear window, while the third would lie down and go to sleep soon after we got underway. After years of skilled observation, extensive analysis, and the unbiased application of pure scientific methodology, I have concluded that we humans are a lot like these dogs.

Some of us prefer to hurtle ahead without knowing where we're headed, some of us care about where we've been, and some are so damn lazy we leave our future up to others. Most are at least somewhat curious about our ancestors, however, but it takes courage to risk discovering that Great Grandpa Zeke was hanged as a horse thief or that innocent little Aunt Daphne made her sizeable fortune in the adult film industry. Even if you have traced your own family tree back to a mere sapling, there is probably a lot about our human ancestry you don't know (yet); and this Chapter tests your commitment to learn more.

NOTE—This Chapter is not intended to be a complete reference source about earlier civilizations, but rather to exercise your skills at learning lessons offered by history and stimulate your curiosity to learn more through your own efforts.

Exercise Guide: In all effective exercise programs, the major elements become progressively more challenging. This Chapter's topic is quite captivating, but to achieve maximum exercise benefits I recommend that you pause and review at the indicated points and take an overnight break at the marked halfway point to absorb new information. Using this approach, you should be able to complete this Chapter within two days.

Like some of you, I endured incredibly boring history classes during which my young brain was force-fed more than I cared to know about Egyptian pharaohs, Greek statues, Roman conquests, medieval plagues, Renaissance artists, and Elizabethan playwrights. This Chapter will prove that history can be a lot more interesting, and that we can learn very valuable lessons about surviving greed, ignorance, and apathy from the footprints and echoes left by our ancestors.

After this inspiring introduction, you probably want to rush down to your corner Archaeology Shop to get a Fedora hat and bull whip and begin digging up your lawn to recover rare relics or collect *coprolites* (a.k.a. fossilized feces). But real archeology isn't as romantic as in the movies, and real archeologists often have:

Dating Difficulties

This is not (necessarily) due to personality problems or body odors, but rather because the dating of objects, events, and civilizations depends a lot upon: the available evidence, the analysis method(s) used, and whether the evidence has been contaminated. Because of these factors, all *prehistoric* and some *historic* dates are estimates, and are usually expressed with statistical margins or error. With that in mind, your first dating experience should be something simple like the:

Earth's Age

In view of the previous section, you probably don't expect an extremely precise answer; but you might be a bit surprised to discover that common replies to the seemingly straightforward question about the Earth's age include:

- A little over 6,000 years old, and

- Roughly 4.5 billion years old (+/–1 percent).

How could the second response *possibly* be 750,000 times larger than the first? The answer is that even though (as you now know) contradictory scientific and philosophical theories can co-exist, they don't always do so peaceably. During the sixteenth century, a French scholar named Palissy was burned at the stake for daring to question the prevailing religious beliefs about the Earth's age. In the seventeenth century, an Archbishop named Ussher laboriously traced biblical events backward in time and proclaimed that the Earth was created in October of 4004BC. This statement served as the foundation for *Young Earth Creationism*, a view based upon literal interpretation of The Bible.

This was soon challenged by researchers who claimed that geological changes simply took far too long to occur within the Creationist's time frame, even though scientific theories about the Earth's age were far from being unanimous. During the 1770s and mid-1800s, two Scotsmen suggested the Earth was a lot older than 6,000 years, and somehow avoided being burned at the stake. In 1897, Lord Kelvin calculated that the Earth was about 24 to 40 million years old, but several years later an Irishman named Joly computed it to be around 90 to 100 million years. Radiometric dating methods have led modern scientists to theorize that the Earth is roughly 4.5 billion years old, plus or minus only 450,000 years.

Young Earth Creationism has also endured, along with variations that include *Creation Science* and *Intelligent Design*. Some Creationists admit the Earth may be a lot older than 6,000 years, but deny that evolution and natural selection are responsible for changing plant and animal life. Because we aren't likely to resolve this centuries-old controversy here, let's move forward at this point to discuss the:

First Folks

The ancient Sumerians believed that the "seed of life" was brought to our solar system by the planet, god, or star they called *Nibiru;* while Creationists believe that according to Genesis: on the third day God created plants, on the fifth day birds and aquatic animals, and on the sixth day other land animals and human beings. Unless some have very seriously misinterpreted the meaning of "day," those who suggest the seeds of life arrived by hitch-hiking on asteroids and meteorites seem to agree more closely with the ancient Sumerians.

Determining when humans appeared remains a moving target for scientists. One popular scientific theory suggests that early human ancestors evolved about 1.8 million years ago. *Homo erectus* bore a striking resemblance to modern humans, but had a more sloping forehead and slightly smaller brain. The taller and more muscular *Homo heidelbergensis* appeared around 600 thousand years ago, and the oldest fossil evidence found of *Homo sapiens* dates back approximately 200 thousand years.

Conversely, Young Earth Creationists insist that all life in its current forms was created a little over 6,000 years ago. Evolutionists say this is not supported by scientific evidence, while Young Earth Creationists respond by arguing that evolution due to random genetic mutations is not feasible even over millions of years. You can decide this for yourself, but in the meantime the rest of us are going to wander into the:

Garden(s) of Eden

Most are familiar with the story of Adam and Eve, the serpent, forbidden fruit, and Garden of Eden; but there is controversy about exactly where the latter may have been. For centuries, the words in various religious documents were interpreted to place it in northern Mesopotamia (modern Iraq), Lebanon, or Missouri (according to the Mormon Prophet Joseph Smith); while other sources claim that the Garden of Eden may have been in Africa or under the current Persian Gulf. When studies of human origins intensified during the twentieth century, the following scientific views emerged to explain fossil and archeological evidence:

- The "Multi-Regional Hypothesis" suggests that human ancestors first spread about the globe, and that modern humans (i.e. Homo sapiens) *subsequently* evolved in several different locations.

- Conversely, the "Out-of-Africa Theory" suggests modern humans evolved in Africa for thousands of years *before* emigrating to the rest of the world.

During the 1980s, DNA testing allowed geneticists to trace lineages backward in time, and their results tended to support the Out-of Africa Theory, We'll discuss this theory in more detail later, but now:

✷ Pause and review what we have covered up to this point.

Early Migrations

In addressing the question of how our ancestors got from wherever they were to where modern humans live today, things get a lot more interesting and a number of topics covered in previous sections of this book (e.g., climate change) begin to fall into place.

Young Earth Creationists claim the animals and modern humans who survived *The Great Flood* on Noah's Ark migrated via frozen oceans and then-exposed land bridges to populate remote parts of the planet. This assumes the sea level was still lower due to a major Ice Age following the flood, but this assumption is contradicted by geological evidence that the current interglacial period began over 11,000 years ago. Some suggest that human migrations took place when current land masses were connected as a supercontinent called *Pangaea*, but at the current rate of sea floor spreading this would have had to occur roughly 200 million years before the Young Earth Creationists say humans were created.

Painting a coherent picture of human migrations using DNA and archaeological evidence is a significant challenge for science and there are huge gaps because most artifacts are from Europe, Africa, and the Middle East, but few from eastern Asia. With this in mind, here is one version about how early humans might have initially spread, based upon the Out of Africa Theory:

Leaving Africa

There is evidence that individuals and bands of "pre-modern humans" left Africa approximately 1.8 million years ago, possibly due to climate changes. Many went to Asia, a few to Europe, and some of the latter became known as *Neanderthals*. For this discussion let us assume that around 200,000 years ago Homo sapiens lived in the vicinity of the

Red Sea. Driven by changing climate in Africa, roughly 60,000 to 90,000 years ago nomadic "band societies" moved into Asia and by at least 45,000 years ago they had reached Indonesia, New Guinea, and Australia by island-hopping on primitive rafts.

These extended family groups or clans seldom had over thirty to fifty members, but collectively may have numbered in the thousands. They were rather loosely organized with only oral traditions, no written languages, and made decisions by consensus or counsel with tribal elders or shamans. Such wandering bands were mainly hunter-gatherers, who followed herds of game and other sources of food. Some of them took coastal routes, ate aquatic animals, and made shell jewelry.

Mass migrations into Europe may have been delayed by its colder climate (the current Ice Age had begun), less abundant food, or a lack of the Neanderthal's hospitality. Around 40,000 years ago, early Homo sapiens who later became known as *Cro-Magnons* immigrated to Europe and drove the Neanderthals into mountainous areas where the latter became extinct roughly 15,000 years later.

Coming to America

Large herbivores, like wooly mammoths, mastodons, and bison, are thought to have migrated across the so-called "Bering Land Bridge," which was exposed between modern day Siberia and Alaska during the last Ice Age when sea level was hundreds of feet lower than it is today. These days there are several main theories about when and how humans first came to the American continents.

The most popular one says they crossed the same land bridge at least 13,000 years ago, then moved southward along the west coast of North America into Meso (a.k.a. Central) and South America, A variation of this theory claims they also took an inland route East of the Rockies; a different one suggests that they followed the West Coast in boats; while yet still another says early humans also came to North America directly from Europe.

✳ **Pause to review the above about *Early Migrations*.**

Between roughly 3000 and 5000 BC, the lifestyles of geographically-separated groups of our restless ancestors almost simultaneously began to change quite dramatically. Those changes gave rise to many famous ancient cultures, and were the first signs of humans:

Becoming Civilized

One way to identify the criteria for being considered "civilized" may be to simply list those attributes undeveloped in most teenage boys. In lieu of this approach however, anthropologists have come up with the following candidates:

1. Irrigation Agriculture
2. Pottery and Metallurgy
3. Creation of City-States
4. Domestication of Animals
5. Construction of Monuments
6. Emergence of Social Classes
7. Religious Temples and Clergy
8. Increased Trade and Commerce
9. Citadels, Palaces, and Bureaucracies
10. Organized Warfare and Building Empires

Please note that dense urban settlements and the existence of written language are not included in this list, because several great ancient civilizations evidently lacked one or both of them. In a later Chapter, we'll discuss suggested reasons why so many wandering bands of humans chose to give up chasing large hairy beasts and settle down at approximately the same time; but now let's focus on where they decided to put down roots. One common factor was a dependable supply of fresh water, because virtually all the earliest civilizations arose near rivers. This is a good point for us to discuss:

Chronology Challenges

Have you ever seen a truly *global* timeline for the rise and fall past civilizations? Are most academics focused so intently on "regional" trees, they have forgotten that they are part of a much larger forest? This is one of very few places where you can find a consolidated chronology of ancient civilizations from around the globe discussed in their (estimated) order of appearance, and this perspective raises some intriguing questions for you to ponder. We'll begin with the:

Cradles of Civilization

Most of us were taught that the Cradle of Civilization lay between the Tigris and Euphrates Rivers in the so-called "Fertile Crescent." But this might not provide a thorough picture, because there is evidence other ancient civilizations emerged at about the same time in Egypt, northwestern India, and on the other side of the globe; but many people have never even heard of some of the following Cradles of Civilization:

> **Mesopotamia**—The strange name of this ancient land does not have anything to do with a passel of potatoes, but instead is Greek for "the land between the rivers." The rivers here were the Tigris and Euphrates, and this is the area of the famous Fertile Crescent. The area's first civilization, called *Sumer*, existed by around 3500BC and was located in the southern region where those rivers converge. The ancient Sumerians established city-states ruled by priest-kings, built temples called *ziggurats* atop tiered platforms, dug

flood control levees and irrigation canals, and created an early form of writing called *cuneiform*.

Today the land that once was Mesopotamia includes: Iraq, northeastern Syria, southeastern Turkey, and southwestern Iran. But our next Cradle of Civilization was on the opposite side of the planet in:

Valdivia—Where? You probably never heard of it, but Valdivia was one of the oldest settled civilizations in the Americas. It thrived by 3500BC on the Santa Elena Peninsula in modern-day Ecuador. Valdivians were mainly farmers and fishermen, but they also hunted. They lived in groups of houses built around central plazas, produced clay pottery and other art objects including so-called "Venus Figurines," traveled on rafts, and traded with others in the Andes and Amazon regions.

But Mesopotamia and South America were by no means the only Cradles of Civilization. At just about the same time, another one began rocking in the:

Indus Valley—Even though we never hear much of it in the western world, by about 3300BC the largest culture in the ancient world was flourishing along the Indus River in what today is northwestern India, Pakistan, and Afghanistan. It was one of the earliest urban civilizations with a peak population of over five million people. One of the most sophisticated cultures of its time, it developed a system of weights and measures and a yet-to-be deciphered written language. Its people were skilled metallurgists, constructed drainage systems, produced handicrafts, and lived in multi-story brick houses.

While archeologists continue to learn more about the Indus Valley civilization, the next one is likely to be more familiar to most people. It arose in:

Ancient Egypt—The ancients actually called it *Kemet* (The Black Land), after the dark, fertile soil of the Nile flood plain. Civilization there coalesced around 3150BC with unification of Upper and Lower Egypt. It was concentrated along the lower Nile in northeastern Africa, and heavily depended upon predictable river flooding, coupled with controlled irrigation of the fertile valley, to yield the surplus crops required to support cultural development and massive building projects. Egypt was ruled by a series of dynasties for almost the next 3,000 years, and its great structures are still considered "wonders of the world."

Although the Ancient Egyptians are the most famous for their pyramids, such structures were by no means unique to their culture. For example, our next ancient civilization also built similar ones on the other side of the globe.

Caral—Also called the *Norte Chico*, this little-known civilization existed on the north-central coast of Peru at around the same time as the Ancient Egyptians. Their government was theocratic. They built platform mounds, amphitheaters, and temples in their major population centers, along with a very sophisticated irrigation system to grow food and cotton in the arid land. Their diet consisted largely of cultivated crops and seafood, and their peaceful culture produced textiles, stone jewelry, musical instruments, and a method of record-keeping that was later adopted by the Incas.

Some include China and Mesoamerica in this list, and we will discuss them later. If you already knew anything about more than two of the above cultures you are smarter than most folks, and if you already knew a lot about more than three of them you are probably a history geek. But most people never even realize that there were also "cradles of civilization" in the western hemisphere.

❋ Pause and review the information on *Cradles of Civilization*.

It is quite interesting that: (1) all these ancient civilizations are believed to have arisen within a few centuries of each other, and (2) none of them is believed to have arisen before about 3500 BC. Most remained relatively close to where they originated, so to further examine how civilization spread we need to look at the following cultures I call the:

Horizon Stretchers

Throughout history, some humans have left familiar surroundings to venture into the unknown. Common reasons include: following their main food sources (e.g., wandering herds), local changes in climate, competition for natural resources (e.g., water), conflicts or persecution, and curiosity about what lay beyond the current horizon. But whatever their motivation(s), some of the ancient cultures that helped spread "civilization" to remote parts of our planet include the:

Minoans—In the late 1800s Sir Arthur Evans, curator of the Oxford University Museum, came across stones in an Athens antique shop that were inscribed with pictographs. He was told they were called "milk stones," and came from the island of Crete where women once used them as charms to ensure they would have enough milk to feed their babies. This led Sir Arthur to discover one of the most important civilizations in the Mediterranean, and one whose influence would be felt for centuries to come.

Who Were They? The Minoan culture was well established by 2700BC, but nobody knows what they called themselves since their name was taken from that of King *Minos* in later Greek mythology. They built magnificent multi-story palaces, created a writing system that served as a foundation for early Greek, were known for their colorful frescoes, pottery, and gold work; and they were the most advanced maritime trading civilization in the region during the time.

Bulls & Minotaurs. King Minos was said to be son of *Zeus* and the princess *Europa*, who Zeus supposedly brought to Crete by transforming himself into a bull and charming her to climb onto his back. Bulls were big with the Minoans and are seen in lots of their art, and there was also the *Minotaur* that lived in the King's basement. This legendary monster, with the head and tail of a bull and a man's body, was supposedly the offspring of a sacred bull and Minos' wife. The Minotaur was imprisoned in a labyrinth beneath the Kings' palace and offered Athenian youths and maidens as sacrifices until it was killed by the Greek hero, *Theseus*.

What Happened to Them? Around 1500BC, the *Thera* volcano erupted on the island of Santorini about seventy miles from Crete, and there is evidence some Minoan cities were struck by tsunami waves. These events led some to suggest the Minoan culture might have been the lost civilization of *Atlantis*.

✻ **Pause and review what you've learned about the *Minoans*.**

The Minoans may have been a great maritime culture, but compared to our next ancient civilization they barely got their feet wet. And if your history classes didn't spend much time on the Minoans, there is a good chance they spent even less time discussing the:

Polynesians—This ancient civilization consisted of people who migrated from southeastern Asia sometime around 2100BC to populate many islands in the so-called "Polynesian Triangle," which has its corners at New Zealand, Easter Island, and Hawaii.

Who Were They? While some groups can be traced back to Melanesians in southeastern Asia, Polynesians are actually a mixed people who also exhibit Indonesian, Malaysian, and Caucasian characteristics. They used the natural signs and possibly celestial objects to navigate across more than 2,000 miles of the open Pacific Ocean to settle upon widely scattered islands.

It Takes A Village. Entire Polynesian villages frequently set out in big double-hulled canoes often not to conquer new territory, but rather to search for new places to live. If the settled islands or island groups were within a day's canoe voyage of each other, their inhabitants typically spoke dialects of the same language; but if not, they often developed in relative isolation with relatively little interest in inter-island communications.

What Happened to Them? By roughly 700AD, many habitable islands in the Polynesian Triangle had been populated, but because they had relatively few gemological or mineral resources, were of little interest to foreign explorers. Their subsequent colonization, however, brought strange diseases, slavers, and missionaries who suppressed Polynesian languages, traditions, and cultural expressions.

✱ **Pause and review what we've covered about** *Polynesians.*

At the same time that Polynesians were spreading civilization across the Pacific, our next ancient culture spread it across the Asian continent. They could be one of the oldest continuous civilizations in the world with unique characteristics, but in school most of us learned relatively little about the:

Ancient Chinese—The uniqueness of the Ancient China's culture is at least partially due to its geography, bounded by: mountains, deserts, steppes, and the sea. Less civilized invaders often adopted Chinese culture and language instead of the opposite, and it is characterized by more cultural and political cohesion and continuity than others of its time.

> **Who Were They?** Agricultural villages were already in Chinese river valleys by 3500BC, and about 2000BC the first dynasties began to emerge. During the *Shang Dynasty* between roughly 1765 and 1050BC, the king and nobles lived in a fortified capital near the Yellow River. They developed a calendar, a writing system, and fielded an army equipped with horse-drawn chariots and bronze weapons. The *Chou Dynasty* came into power around 1050BC and created a feudal system in which royal family members and military leaders were given land in exchange for their support. Chou rulers also introduced the "Mandate of Heaven," which claimed the gods gave them approval to reign.

> In about 220BC, the Chou were replaced by the *Qin (or Ch'in) Dynasty*, which ended the feudal period and marked China's unification under a strong central government, further expansion of its empire, and emergence of philosophies like Confucianism and Taoism. The Qin established standard weights and measures, created a legal system, minted coins, and undertook major construction projects, including:

> **Great Walls & Mud Armies**. The first Qin Dynasty Emperor built a network of highways and began connecting existing walls to complete the *Great Wall of China*, but a million workers died working on them. This did not particularly endear the Emperor to the people, and he evidently got a bit nervous about what might happen to him in the afterlife. In 1974, local farmers

discovered a necropolis including offices, hallways, and stables in his burial mound, along with an entire army made of terra cotta clay. It included: over 8,000 life-size soldiers, 130 chariots harnessed to 520 horses, 50 cavalry horses, figures of officials, strongmen, acrobats, and musicians; as well as many weapons.

What Happened to Them? A General from peasant stock overthrew the Qin in 202BC to found the *Han Dynasty*, during which China's empire expanded into Indochina, Manchuria, Korea, and central Asia. They made porcelain and paper, opened the famous Silk Road which made trade possible with cultures in the Mediterranean as far away as 4,000 miles, and established sea trade with Indian civilizations that prized Chinese silk, jade, and other goods. The Han Dynasty lasted until around 220AD, which is roughly the same time frame as that of the Roman Empire.

✳ **Pause and review the above about the *Ancient Chinese*.**

While the Polynesians were paddling across the Pacific and the Ancient Chinese were developing dynasties, our next group of ancestors was one of the first to spread civilization in the Americas.

Olmecs—Their name meant "rubber people" in the language spoken by the Aztecs. This wasn't because the Olmecs were especially limber, but because they gathered latex from rubber trees. Their civilization arose around 1400BC, and was centered in the lowlands of south-central Mexico in the modern states of Tabasco and Veracruz. There rich soil and the nearby Coatzacoalcos River supported a dense population which included an elite social class that enjoyed recreational activities, cultural advances, and luxury items.

Who Were They? Similarities in physical characteristics, religious symbols, language, hair styles, and body scarification practices have suggested the Olmecs may be traced back to Melanesians or groups in West Africa. They practiced ritual bloodletting and introduced a ball game that became popular in later Mesoamerican cultures. These games took the concept of elimination tournaments to an whole new level, however, because there is evidence that losing competitors might have sometimes been sacrificed.

Jade Trade & The Big Heads. This is not the name of a rock band, but the descriptor of several things for which the Olmec's elite social class was well known. They traded for semi-precious stones like obsidian and green stone (i.e., jade) used to create jewelry and other art objects. They developed the first writing system in Mesoamerica and a calendar, built pyramids, and also carved "Colossal Heads" that appear to be wearing helmets with ear flaps.

What Happened to Them? The Olmecs established a number of villages and several powerful urban centers, but the latter were abandoned between about 900 and 400BC. During the next fifty years, the Olmec culture became virtually extinct, some say, due to climate changes, volcanic eruptions, shifts in paths of rivers, or invasion. They may have been gone, but their influence lived on in other Mesoamerican cultures including the Mayas and Aztecs.

✳ Pause and review what you've learned about the *Olmecs.*

At about the same time that the Olmecs were playing their bloody ball games and carving colossal heads in Mexico, another ancient civilization emerged back in the so-called "Old World." They were called the:

Mycenaeans—Our next Bronze Age culture again takes us thousands of miles back into the eastern Mediterranean and onto the Greek mainland, where the Mycenaeans appeared shortly after the Minoan civilization collapsed.

Who Were They? Unlike the Minoan culture, which was based upon trading, the Mycenaeans were dominated by a warrior aristocracy that extended its influence through conquests. By 1400BC, they had defeated the Minoans and adapted their form of writing; and according to Greek legend, also conquered the city-state of Troy. Mycenaean bronze swords have been found as far away as the Caucasus, and their double axes have been discovered in the British Isles.

Armed Mummies in Beehives? The Mycenaeans often mummified their dead nobles and buried them in a sitting position within high-vaulted, circular tombs shaped like beehives. They were also frequently interred with jeweled weapons, armor, and other military equipment, just in case conflicts had to be settled in the afterlife.

What Happened to Them? The Mycenaean civilization collapsed in around 1100BC. Numerous cities were sacked, some of the population fled to Cyprus and other islands, bureaucracies folded, literacy took a major step backward, and Greece entered into what many considered to be a "dark age." Historians traditionally blamed these events upon invasions or uprising by another group called the *Dorians*, but others claim the Mycenaean's rapid decline was due to a series of natural disasters, like earthquakes and widespread droughts.

✱ **Pause to review what we've covered about the** *Mycenaeans.*

Until now, the great ancient maritime cultures have included the Minoans and the Polynesians, but our next one is probably the most famous. They are the:

Phoenicians—At roughly the same time the Mycenaean culture was in its decline, the Phoenician civilization was emerging along the coastal region of the Fertile Crescent. In the centuries after 1200BC, they became the greatest maritime trading power of the time using sturdy vessels they called *galleys*.

Who Were They? They may have referred to themselves as *Canaanites* and might not have even viewed themselves as a single ethnicity. Their city-states were politically independent, but worked together to form alliances, if needed. The Phoenician culture was the first to use an alphabet at the state level, and their extensive voyaging and trading spread usage of their alphabet to Africa, Europe, and the Greeks who (after adding the vowels) passed it along to the *Etruscans,* who passed it on to the Romans, etc., etc.

Snail-Based Trading. Some say this culture's name came from the ancient Greek word *"phoínios"* which means purple, because the Phoenicians based a large portion of their trade on the powdered Tyrian Purple dye made from shells of sea snails and used to color textiles. However, they also traded in: wood (including cedar trees used for Solomon's temple in Jerusalem), slaves, glass, hunting dogs, silver, gold, copper, and wine. The Phoenicians are said to have circumnavigated Africa and did business as far off as Spain, Ireland, and England. On coins minted in their outpost of Carthage in northern Africa, there is also a map that some people say depicts America off to the west.

What Happened to Them? The Phoenician cities and commercial outposts eventually were overrun by more powerful military cultures. Their city of Tyre fell to the Babylonian *King Nebuchadnezzar* (what a great name!) in 573BC, and again to the famous Macedonian warrior *Alexander the Great* in 332BC. Their outpost of Carthage was sacked by the Romans in 146BC, and their former cities of Tyre and Sidon are now located in modern day Lebanon.

✴ **Pause here and review the above about the *Phoenicians*.**

Ancient Greeks—Greece recovered from its "dark age" around 750BC, and by roughly 530AD the Greek city-states had become famous for their contributions to architecture, philosophy, dramatic arts, mathematics, medicine, and science and they shared trading in the Mediterranean with the Phoenicians. Since so much has already been written and taught about the Ancient Greeks, I won't dwell upon them here, but you really should know the story about how the Trojans learned to beware of Greeks bearing gifts.

Ancient Romans are included because according to legend, a pair of twins who were raised by a female wolf turned a relatively small Italian agricultural community into one of the ancient world's largest empires. From its origin as a monarchy in roughly 750BC, Ancient Rome initially evolved into an aristocratic republic and then into an increasingly autocratic empire. Through conquests and assimilations, the Roman Empire extended its dominance into northern Africa, Asia Minor, the Balkans, and southern, eastern, and western Europe. But because so much has already been studied and written about this early superpower, let's move on.

✽ **Congratulations! You have reached the mid-point of this Chapter. Take an overnight break to absorb the information covered up to this point.**

The ancient ancestors we discussed in the last section were greatly responsible for extending civilization to remote parts of the globe. As we approach modern times, one might expect the information about past cultures to become more familiar, but this is not the case as you will discover from this next group of:

Ancient Americans

After spending a lot of time learning about ancient Mediterranean and European cultures, I was quite surprised to discover that some of our most interesting and controversial ancestors lived in the Americas. So in this section, we are going to explore the fascinating facts, captivating legends, and important contributions of cultures with poorer publicity departments and/or less effective visitors' bureaus. For example, how much do you really know about the:

Zapotecs—At around the same time folks in the Mediterranean were beating each other up, the Zapotec civilization arose in the *Oaxaca Valley,* about 125 miles from modern-day Mexico City.

Who Were They? One legend claims the Zapotecs arose from rocks, trees, and animals; while another says they descended from supernatural beings who lived among the clouds. The valley's natural branches were initially each home to competing societies separated by a fifty-mile wide "no man's land," but in roughly 500BC several large settlements lost most of their populations while one sprang up atop a mountain in the former no man's land. This new fortified city of *Monte Albán* had: step pyramids, tombs, ball game courts, houses, and irrigation canals which enabled them to cultivate crops in the surrounding valley, and at its peak more than 25,000 residents.

Don't Use That Tone With Me! Monte Albán's rise marked the beginning of conquests that extended Zapotec influence well beyond the Oaxaca Valley. The "Conquest Slab," discovered in its main plaza, depicts the heads of more than forty conquered states, but the Zapotecs were much more than warriors. They developed a calendar and a writing system that were predecessors of those used by the Mayas and Aztecs; and spoke a tonal language in which meanings of words depended upon whether the pitch of a speaker's voice was high, low, rising, or falling.

What Happened to Them? Although the area was subsequently occupied by others including the Aztecs and Spaniards, descendents of the Zapotecs still make up the largest portion of the modern-day Mexican State of Oaxaca's population. Their weavers are known for their fine quality textiles, while their artisans are famous for gold and silver jewelry, pottery, and leather goods.

✱ **Pause and review what you've learned about the Zapotecs.**

The Zapotecs were by no means the only ancient people in Mesoamerica, and now that you know more about them you are prepared to explore the even more mysterious and fascinating culture of:

Teotihuacán—People had lived in the area roughly thirty miles northeast of what is today Mexico City for centuries by around 150BC, when some began building what would become one of the most impressive and the largest urban centers in the Americas. Its original name, purpose, and inhabitants may be all shrouded in mystery, but its heritage can still be seen throughout Mesoamerica.

Who Were They? Nobody knows who its founders actually were. The name Teotihuacán was later given to it by the Aztecs, and means "The Birthplace of the Gods." The city's peak population included more than 150,000 residents, who lived in multi-family compounds and multi-story apartments with plaster walls decorated with elaborate murals. Teotihuacán's great step pyramids, platform temples, and the city itself appeared to be aligned with celestial objects; and while Zapotec influences are apparent, there is evidence that Teotihuacán might have been a multi-ethnic city.

Demanding Deities. The Teotihuacáns worshipped some of the same gods as other Mesoamerican cultures, including the Feathered Serpent the Aztecs called *Quetzalcoatl*, who was said to appear in human form as a light-skinned man with a beard. Like the Mayas, they believed that the Universe had to be periodically renewed if the gods deemed it worthy and that human sacrifices could help convince the gods to do so. The primary goal of the Teotihuacán warriors wasn't territorial conquest, but instead to capture prisoners for their human sacrifices. Victims were brought into the city to be decapitated, have their hearts cut out, bludgeoned to death, or buried alive.

What Happened to Them? Teotihuacán continued to be built until around 250AD and reached its peak about 450AD, by which time the Teotihuacáns had extended their power as far as Guatemala and the Yucatan peninsula. Its population began to decline in the sixth century due, some say, to droughts and resulting famine. This led to internal unrest, which in turn resulted in the burning of major civil structures and homes of the elite class, and destruction of many statues. Inhabitants abandoned the city by about 700AD, and after 1320AD Teotihuacán became a pilgrimage site for Aztecs who believed the gods had gathered there to create the Sun and Moon. The Conquistador Cortés came across its ruins in 1520AD, and today it is the most visited archaeological site in Mexico.

✱ **Pause here and review what we've covered about *Teotihuacán*.**

While the ancient Greeks and Romans were busy philosophizing and gladiating around in the Mediterranean, one of the most enigmatic cultures was emerging in South America. Welcome to the world of the:

Nazcas—The first century AD was not a terrific period for the Roman Empire. Christ was crucified, Rome burned, and Mount Vesuvius erupted. But back in the Americas, on the southern coast of modern day Peru in the Ica Valley, and along the Rio Grande River, the emergence of the Nazca civilization was well underway.

Who Were They? The Nazcas were one of the cultures collectively called the Andean Civilizations. They were heavily influenced by the preceding *Paracas* culture, and they contributed to subsequent ones, like the *Incas*. Although the Nazcas prevailed for less than 1,000 years, they left behind several mysteries that have puzzled us for many generations.

Jars & Skulls The Nazcas created beautiful textiles and ceramics, and built an impressive system of underground aqueducts to irrigate their crops; but their burial practices were a rather unusual. Nazca graves included: bundles of bones, stacks of skulls, and decapitated skeletons where missing heads had been replaced by ceramic jars with painted-on human faces.

There are severed skulls that some claim were "trophy heads," displayed to indicate a leader's prestige.

The Nazcas sometimes bound cushions to the front and boards to the back of their infants' heads to create elongated skulls. Ancient alien advocates claim this was to make them resemble early extraterrestrials, but others say it was done to denote social status. There is also evidence the Nazcas performed a primitive form of skull surgery called trephination to reduce pressure on the brain or perhaps religious purposes, and signs of healing indicate some of these individuals even survived!

The Nazca Lines. Discovered in 1939, these are the most famous mystery about the Nazca culture. They include geometric shapes and animal figures, plus one called "The Astronaut." They were created by removing the surface layer from the underlying soil and have endured for centuries because of the desert environment. Their purpose has been widely debated and the theories include: some sort of calendar, figures for gods to admire, ceremonial paths, and extraterrestrial landing strips. None of these have been proven, yet.

What Happened to Them? The Nazca culture began to decline in roughly 500AD, and had virtually collapsed by 750AD. Some say this resulted from effects of *El Niño*, which included flooding and erosion. During the 1500s, they were virtually wiped out by the Spanish Conquistadors, and a lot of fascinating history was lost forever.

✱ **Pause and review what you've learned about the** *Nazcas.*

Our next ancient civilization is just as intriguing as the Nazcas, but to learn more about them we need to travel thousands of miles and onto the North American continent to the land of the:

Anasazis—The first humans in the area currently known as the *Four Corners*, where today's states of Arizona, Utah, New Mexico, and Colorado meet were hunter-gatherers; but around 100AD, groups of farmers began to settle in the region. These farmers didn't always get along so well with their more nomadic neighbors, which may be why archeologists began calling them the *Anasazi*, a later Navajo term that means "ancestors of the enemy." Their descendents don't like the name, but as yet cannot agree upon an acceptable substitute.

Who Were They? We don't know what they called themselves because they left no written records, but they evidently changed their lifestyle several times. From about 100AD to 500AD, the people lived in caves or shallow pit houses, gathered wild vegetables, cultivated corn, hunted small game animals, and stored their food in covered pits. During the next 250 years, they moved into relatively deep subterranean caves or atop mesas, grew bean and squash crops, and domesticated turkeys, which were used for feathers, pets, and to keep bugs out of gardens. From around 750AD until 1150AD they moved above ground and constructed a number of villages and large communities with many adjoining rooms and underground chambers, but they were not finished changing lifestyles yet. Over the next 150 years or so they built cliff dwellings under ledges and in sheltered recesses of canyon faces and mesa walls. Their people moved from the smaller villages into larger communities, many of which were in defensive locations and included ropes or ladders that could be raised in case of attack.

Kivas & Kachinas. During inclement weather, the Anasazis took shelter in covered pit houses or in rooms called *kivas*. Their early kivas were typically underground chambers, while later ones were built on the surface. In larger communities, it wasn't unusual to have a kiva for every five or six residences, and they began to be used more for ceremonial purposes. Most were round or square, with holes in their floors that represented places where ancestors emerged from the underworld, and entrance holes in their roofs which let at least some smoke escape. Kivas continue to dot the American Southwest.

Kivas are closely tied to the *Kachina* belief system, which maintains that all things have both physical and spiritual forms. Kachinas represent locations, elements, qualities, natural phenomena, concepts, ancestors, etc.; and are spirit beings who can use their particular power to benefit humans, if they are venerated and respected. Anasazi legends tell of "Star People" who brought them knowledge, and some of their descendants call kivas "Cloud Houses."

What Happened to Them? Anasazi communities in the Four Corners region were essentially abandoned by around 1300AD due to drought, crop failures, and/ or conflicts with ancestral *Navajos* and *Apaches*. The people moved and established new communities where the water supply was more reliable; and the Spanish called them Pueblo People, because by then many lived in small villages. But when the Spaniards tried to Christianize and collect tribute from them they revolted and remained free for the next fourteen years. By the eighteenth century, diseases, and violence had reduced their number of settlements to approximately thirty; but in the early twenty-first century, their descendents among the *Hopi*, *Zuni*, *Acosta*, and *Laguna* people numbered approximately 75,000.

✳ **Pause and review the above information about the**
   Anasazis.

We will return later to learn more about other ancient North
Americans, but now it's time for us to head back South once
again to visit the mysterious:

Mayas—All many people know about this ancient
Mesoamerican culture is that they developed some sort of
calendar, which was said to predict the end of the world.
But there are many other fascinating things to learn about
the Mayas.

Who Were They? The Mayan civilization arose in
roughly 250AD, and was heavily influenced by the
Olmecs. Within the next 850 years, their numbers
increased to an estimated two million; with over forty
major Maya cities spread across modern-day southern
Mexico, Guatemala, and Belize. Interestingly, today's
name for the Yucatan peninsula came from the Mayan
word *uicathan*, which was how they replied when the
Spaniards asked them what they called their land.
It actually means "What do you mean? We don't
understand you."

They enjoyed dramatic performances and dancing, but
waged war on other tribes and rival Mayan cities, and
they played their version of the ball game introduced
by the Olmecs. In it, teams tried to maneuver a ball
through stone rings high on the sides of the court, and
it represented the struggle between light and darkness.
Losing team members were sometimes sacrificed, but
it could be worse. Captured rulers were often tortured
for up to a year before being forced to play the game
then decapitated, whether they won or lost.

Religious Sacrifices. The Mayas believed ritual bloodletting and sacrifices were ways to contact their gods

and ensure fertility. After (sometimes) being painted blue, victims had their hearts cut out atop pyramids; and members of royalty were not exempt from religious sacrificing. As intermediaries between their gods and people, nobles drew their own blood by jabbing stingray spines through their ears or genitals, and pulling thorn-studded cords through their tongues. But even these actions didn't get them into Mayan heaven, which was reserved for those who had been sacrificed or had died in childbirth.

One of their most important gods was the Feathered Serpent they called *Kukulcán*, who was later adopted as *Quetzalcoatl* by the Toltecs and Aztecs. While the Mayas were fairly short with brown eyes and dark complexions, in his human form Kukulcán was said to be tall with blue eyes and light skin.

Doomsday Calendar? The Mayas actually had several calendars including a sacred one used for religious purposes and naming children, which consisted of thirteen months with twenty days in each; and one used for planting, which consisted of eighteen months plus an additional five-day period. Their so-called "long count" began in 3114BC and measured time in *baktuns,* which are about 394 years long. The number thirteen was sacred to the Mayas, and the thirteenth baktun ended on December 21, 2012. Some people claimed this signaled the end of the world, while others said it simply designated the start of the next cycle. Since you are reading this book after that date, it is probably prudent to side with the latter interpretation.

What Happened to Them? Most southern Mayan cities were abandoned sometime after 900AD, but ones on the Yucatan peninsula existed until the early 1,500s. By the time the Conquistadors arrived, most residents lived in farming villages. The Mayas were the first Mesoamerican civilization with a fully-developed written language, but nearly all their records, called *codices,* were destroyed by the Spaniards. Sizeable populations of their descendents still live in southern Mexico, Guatemala, and Belize; and Mayan influence is evident as far away as El Salvador and the Honduras. Their languages are still spoken, and many of their descendants still practice a religion that is an odd combination of Roman Catholicism and Mayan cosmology and rituals.

✱ Pause and review what you've learned about the Mayas.

Our next culture is just as fascinating as the Mayas and a lot more controversial. But to learn more about them, we need to get our feet wet and venture offshore into the South Pacific Ocean to visit the:

Easter Islanders—On Easter Sunday in 1722, a Dutch explorer named Jacob Roggeveen discovered an uncharted speck of land and named it Easter Island, although these days it is also known as *Rapa Nui.* It is about 15 miles long, 7 miles wide, and nearly 1,290 miles from its closest inhabited neighbor, Pitcairn Island. It is the southeastern corner of the Polynesian Triangle, and its original name is just one of many controversies about the ancient culture on this tiny volcanic island over 2,200 miles off the coast of modern-day Chile.

Who Were They? Many believe the first Easter Islanders were Polynesians who came by canoe from the Marquesas or Gambier Islands between 300AD and 1200AD, and this theory is supported by DNA evidence. Others point to things they say indicate some were from South America, like certain plants (e.g., sweet potatoes) which were staples of early South American diets, and stonework strangely similar to that of mainland cultures (e.g., the Incas). And there is also the Polynesian name for a tiny islet to the East that means "bird's island on the way to a faraway land," which seems to indicate that existence of South America was known before Roggeveen's discovery.

Mysterious Moai. The first settlers found an island with giant palms and a sizeable number of native birds, and they initially lived in strangely-shaped houses some claim were derived from overturned boats. They existed for the most part in peace as clans, and worshipped their ancestors. A key element of their religion involved the construction of giant stone heads called *moai,* placed around the island facing villages so the ancestors they represented could watch over their people. But it appears this same devotion might have contributed to the downfall of the island's culture.

Easter Island's population may once have numbered in the tens of thousands, and they cut down the palm trees to build homes, boats, and use to move the giant moai from the quarry to their final locations. But erosion of the bare land washed away the topsoil needed to grow crops, and without boats they could not catch enough fish to support their large population. The native birds were decimated by human and rodent predators competing for increasingly scant resources, hostility arose among the clans, many of their sacred moai were toppled or destroyed, and some claim that cannibalism even occurred.

The Bird Man Cult. After teetering on the brink of extinction, the thing that saved Easter Island's society came from an unusual direction. The so-called "Bird Man Cult" arose around the previously unexceptional god *Makemake,* and became the principal religion. It was centered in the village of *Orongo,* where annually male representatives from each clan raced down the steep cliffs and swam to the islet of *Motu Nui* to find that season's first egg of the *Manutara* bird. The first competitor to climb back up the cliffs to Orongo with an intact egg won control over the distribution of island resources by his clan for the next year. This practice was the islander's version of term limits.

What Happened to Them? Roggeveen estimated there were roughly 2,000 to 3,000 inhabitants in 1722, and this increased to around 4,000 by the early 1800s. But the arrival of foreigners included slavers, who took islanders to work in South American mines and introduced unfamiliar diseases such as smallpox and tuberculosis; and by 1877, the island's native population had dropped to 111. European missionaries and entrepreneurs began buying up land of natives who died, and the island was sold to Chile in 1888. Today the only trees on the island are planted, and curious visitors fly in from Chile to land on the extended runway that once served as an emergency landing strip for the U.S. Space Shuttle.

✱ **Pause and review what we've covered about *Easter Islanders*.**

Around the same time that Easter Islanders were pillaging their palm trees and making a mess of moai, back on dry land another ancient culture was emerging which some say borders upon the mythical. Welcome now to the land of the:

Toltecs—A substantial portion of what is known about the Toltecs was derived from writings of the Aztecs, who considered the former to be their cultural and intellectual predecessors. Skeptics argue that such accounts are tainted by the tendency to glorify the "good old days," and are likely a mix of fact and fiction.

Who Were They? Around 750AD, a group of people who spoke the Nahua language founded a city they called *Tula* in central Mexico, about forty-five miles from modern-day Mexico City. The name Toltec has several meanings including: "master craftsman" and "urban dweller"; but whoever they were, the Toltecs were heavily influenced by the Olmecs and had links to both the Zapotecs and Mayas. They smelted metals, produced advanced stonework, and had considerable knowledge of astronomy. They drank the fermented beverage called *pulque*, their religious ceremonies included Sun worship and human sacrifices, and they played the sacred ball game they called *tlatchli*.

Tula was the Toltec capital, and between 900 and 1100AD, it had roughly 40,000 to 60,000 residents. It had paved streets and its ceremonial center was surrounded by pyramids, clusters of homes, and temples guarded by statues of fierce warriors. Atop the main pyramid were stone statues of gods called the *"Telemons of Tula,"* which also supported a temple's roof. Erich von Däniken, author of *Chariots of the Gods*, suggested they represented beings from Atlantis or aliens and that some had ray guns strapped onto their waists.

Warriors with Wooden Swords? The culture in Tula became increasingly militaristic, war and death became dominant themes of the city's stonework, and the Toltecs' practice of human sacrifices most likely began there. They conquered the great city of Teotihuacán around 900AD, and between then and about 1100AD, their empire encompassed most of central Mexico, the Gulf Coast and Yucatan peninsula, and maybe even the Pacific Coast and Chiapas. There was a myth that Toltec warriors carried wooden swords so they wouldn't kill their enemies, but their swords had razor-sharp obsidian pieces embedded in those wooden blades and were said to be capable of decapitating horses. Perhaps it would be wise to re-consider that myth!

What Happened to Them? There is archaeological evidence that Tula had lost much of its former power and was at least partially abandoned by about 1200AD due, many believe, to a drought and resulting famines. The Toltecs mysteriously disappeared in the twelfth century. Their great city's temples and pyramids were razed by other tribes; and Tula was laid to waste probably by the *Chichimecas*, whose primitive culture was later described as people who "burrow in caves or at best live in cabins of straw."

* **Pause and review what you've learned about the *Toltecs*.**

Not long after the Toltecs were settling down in central Mexico, another ancient civilization was emerging in what today is the southeastern United States. They would become known as the:

Mississippians—Theirs was the last major prehistoric civilizations in North America, and one of its most important, because it established the ancestral foundation for a number of later Native Americans. Despite this, however, the Mississippians rarely get the attention they deserve in most history classes.

Who Were They? Around 800AD, some groups of people began to abandon the nomadic hunter-gatherer lifestyle, settle in communities, and rely more on cultivating crops in the rich soil of river flood plains. Their earliest settlements were in Mississippi, Louisiana, Alabama, Georgia, and the Florida panhandle; but spread northward into Arkansas, Missouri, Illinois, Tennessee, Kentucky, the Ohio River Valley, and Indiana, with extensions as far north as Wisconsin and Minnesota and west to the Great Plains. By 1200AD, the Mississippians had built major population centers and ceremonial complexes, and one of the largest was at *Cahokia* near what is today Collinsville, Illinois.

Mound Towns. Stepped pyramids, platform temples, and central ceremonial plazas were relatively common in Mesoamerica, but such structures were not found in North America until the Mississippian culture. Major settlements had one or more earthen mounds often topped by a chief's residence or a temple, and near a central plaza. The largest of these mounds was Monks Mound at Cahokia, which has four stepped terrace levels, and is approximately 1,000 feet long, 700 feet wide, and 100 feet high.

Major population centers were ruled by priest-chiefs, who controlled specific territories and were responsible for the distribution of food among the urban centers and outlying communities. Mississippian culture was based upon the cultivation of corn, beans, squash, and other crops; and while they had no written language or stone architecture, they carved, molded, and engraved craftwork in shell, stone, copper, wood, and clay. They traded via a network that extended as far as the Gulf of Mexico, Atlantic coast, Great Lakes, and the Rockies; and played a game they called *chunkey*, which included rolling stone discs down prepared courts. Warfare was not uncommon, and often resulted in confederacies and alliances. But in case those failed, their major settlements were sometimes surrounded by ditches, earthen ramparts, or stockades.

What Happened to Them? The Mississippian civilization had already begun to decline by the time European explorers came into the area. Residents left Cahokia between 1350 and 1400AD, possibly moving to other rising centers. There was decline in mound-building and ceremonialism, political turmoil and warfare increased, and more defensive structures were constructed. By about 1500AD, much of their population had dispersed or was under serious stress, and some claim this was due to the climate change during the Little Ice Age.

Encounters with the Spanish led by Hernando de Soto in the early 1500s left many Spaniards and Native Americans dead, and the introduction of diseases like measles and smallpox also took their toll on the Mississippian chiefdoms. Some of their people used horses introduced by the Europeans to return to a nomadic life, a few maintained oral traditions, and some didn't even know that their ancestors built the mounds dotting the land. But the Mississippians were likely ancestors to many Native American people in the region, including the: *Creek*, *Cherokee*, *Chickasaw*, *Choctaw*, *Missouri*, *Natchez*, *Osage*, *Seminole*, and others.

✳ **Pause and review what we've covered about *Mississippians*.**

Even if you didn't spend much time learning about the Mississippians, you have probably at least heard of our next group of ancient American ancestors. They are the:

Aztecs—The name Aztec actually means "the people from *Aztlán*." It does not refer to any particular ethnic group, but instead to several groups that spoke the Nahua language and claimed heritage to a mysterious place they called Aztlán.

Who Were They? During the sixth century AD, nomadic hunter-gatherers migrated into central Mexico from a place in the north called Aztlán, which some claim may have been as far away as the southwestern United States. They called themselves *Mexicas*, and at first were outcasts in the new land dominated by powerful city-states. According to legend, the Mexicas were shown a vision of an eagle, perched on a cactus and eating a snake, which somehow told them to make their home on a swampy islet in Lake *Texcoco*. This unlikely site would become *Tenochtitlan* (modern-day Mexico City), one of the largest cities in the western world and the capital of the most influential empire in Mesoamerica.

The Mexicas began to consolidate their power and form key alliances. One of their most important was so-called "Aztec Triple Alliance" which, after rising to power around 1428AD, became more commonly known as the Aztec Empire. It would dominate the Valley of Mexico, and extend its influence to the Gulf and Pacific coasts and as far as Guatemala. But it was unlike other empires because it gained and held power not only through military conquests but by: installing friendly regimes in conquered lands, arranging marriages between ruling dynasties, making them dependent upon the central government for luxury items, and collecting tribute from outlying city-states.

Bargains in Beans. Aztec tribute included luxury items like green stone (i.e., jade) beads, adorned clothing, and exotic feathers; plus practical things like food, firewood, and textiles. Nobles owned the land, but common people had access to farmland under arrangements that ranged from sharecropping to slavery. Without draft animals, trade was done on foot along routes with rest stops approximately every six to nine miles, which were used by runners who carried messages and made sure the inland nobles had fresh seafood. Small purchases were made using cacao beans. A turkey egg cost around 3 beans, a small rabbit cost about 30 beans, gold figurines cost approximately 250 beans, and fathers could sell their daughters to be concubines or religious sacrifices for roughly 500 to 700 beans.

Aztec Life. The city of Tenochtitlan was laid out symmetrically, had canals for transportation, and was centered around a great pyramid that rose 164 feet above a ritual area. The Aztecs attended poetry contests, dramas, and other events that often featured musicians and acrobats; and played a variant of the ball game they called *Tlachtli*, in which players attempted to maneuver a ball through stone rings using only their knees, hips, and elbows.

The Aztecs shared some common gods with other Mesoamerican cultures, and this eclectic philosophy resulted in unusual combinations of beliefs and practices. They believed that their carved stone figures and statues were the physical representations of supernatural forces; their god Quetzalcoatl was also said to be light-skinned man with a beard; and they practiced human sacrifices. Estimates about the number of their sacrificial victims vary greatly, but some apparently considered it to be quite an honor. In one account, an enemy warrior who had been freed for his bravery returned voluntarily to be sacrificed.

What Happened to Them? The Aztec Empire had reached its peak before arrival of Conquistadors led by Cortés in 1519AD. A smallpox outbreak killed almost half of Tenochtitlan's population in 1521AD, and by aligning with rival tribes the Spanish toppled the Aztec Empire that very same year. Some claim however, the Aztec's defeat was facilitated by their emperor Montezuma's belief that Cortés was their returning light-skinned god Quetzalcoatl.

Nahua dialects are still spoken today by approximately 1.5 million people in the mountain areas of central Mexico, and modern Mexican cuisine is heavily influenced by the Aztecs. Their words, like tomato and chocolate, have found their way into the English language; and Mexico's national flag still bears the symbols of the Mexica's ancient vision. The Aztecs are still with us today.

* **Pause and review what you've learned about the Aztecs.**

Compared to certain other ancestral groups, our next culture did not produce a plethora of pyramids, but they did make a mess of mummies. But to learn more about them, we need to go back to the highlands of Peru and visit the intriguing:

Incas—In their native Quechua language the word *Inka* means "ruler" or "lord," but the Spanish used the term to refer to all members of this culture instead of just its ruling class. The Incan capital was in Cusco in the highlands of modern-day Peru, and their empire was the largest in pre-Columbian America. Their leaders were called *Sapa Inkas* or "ultimate rulers," and were thought to be direct descendents of *Inti*, their Sun God.

Who Were They? According to legend, the first Sapa Inka *Manko Kapak* and his sister-wife *Mama Okyo*, were directed by *Inti* to find a place for his people to dwell and teach them civilized ways. Manko wandered the Andes, plunging a golden staff into the earth until he came to a fertile valley where it vanished into the ground. There he founded Cusco, which means "navel of the world." In the twelfth century, the Incas were basically a pastoral tribe, but in 1438AD they undertook a massive period of expansion under a Sapa Inka whose name *Pachakuti* literally meant, "World Shaker."

Spies & Bribes. From 1438 to 1533AD the Incas used a variety of methods including conquests, peaceful assimilations, and bribery to build an empire that would encompass Peru, a large portion of modern day Ecuador, south central and western Bolivia, northern and central Chile, southern Columbia, and northwestern and central Chile. Pachakuti sent spies into the regions he desired to gather intelligence about their wealth, military might, and political organization. He next sent messengers or even visited the leaders himself to: extol benefits of joining the empire, offer them gifts (including concubines), and promise them riches as regional rulers for the Incas. Many acquiesced peacefully, but those who didn't were visited by the mighty Incan army.

Labor, Lice, & Taxes. Besides performing public service work called *mita*, all individuals were required to pay taxes; but the Incan tax system was based upon goods instead of money. Farmers kept one-third of what they produced, and turned two-thirds over to the government. People who were too poor to pay had to give the Incan tax collectors "a large quill full of live lice" every four months just so they wouldn't forget their responsibilities. The Incas didn't have a written language, but kept accurate records using devices they called "*khipus*," which consisted of pieces of knotted strings.

Religious Beliefs. Most of their deities were gods of nature, but the Incas also worshipped their ancestors and mummified many of them. Their royal mummies were lovingly tended, dressed in fine textiles, and taken to visit each other and attend celebrations. The Incas believed in reincarnation and that death was followed by a difficult passage to the next world, during which one's spirit was led by a dog that could see in the dark. Individuals who had obeyed the Incan moral code of "don't steal, don't lie, don't be lazy," went on to live among flower-covered fields and snow-capped mountains in the Sun's warmth, while all others spent their eternity in the cold earth eating stones.

According to Incan legend, the god *Virakocha* taught their ancestors about agriculture, astronomy, mathematics, technical skills, and the civilized ways needed to advance their culture. He was a tall, light-skinned man with green eyes and a beard, he dressed in a long robe and sandals, wore the Sun like a crown, spent time living among the people, and walked upon the water. He left one day, but promised to return again. Does any of this sound familiar?

What Happened to Them? Even before arrival of the Spaniards in 1526AD, smallpox spread over the Inca's efficient road system and within a few years claimed 60 to 90 percent of their population. Weakened by disease and civil unrest, the Incas were no match for Conquistadors led by Francisco Pizzaro. The current Sapa Inka was confused by Pizzaro's Surrender Agreement and taken hostage. Even though he offered his captors enough gold and silver to fill his prison cell, after receiving the ransom they still executed him. The last Incan stronghold fell in 1572AD, and the Spaniards used the Inca's own mita system to work the people to death in the gold and silver mines. Every family had to provide one member to work in the mines, and when they died (often within a year or two) their family was required to provide a replacement.

As with other American cultures they conquered, the Spanish destroyed many Incan temples and palaces and replaced them with their own cathedrals and colonial structures. And since the Incas evidently didn't leave written records, much of what we know about them is based on later "interpretations" by foreigners with their own agendas and perspectives.

✳ **Pause and review what we've covered about the *Incas*.**

To Be Continued

As proof that Chapter topics were not chosen arbitrarily, it is worthwhile repeating that there is evidence climate changes might have played roles in early migration patterns as well as the development/collapse of a number of ancient civilizations like those of the Olmecs, Mycenaeans, Teotihuacáns, Nazcas, and Toltecs. It is also worthy to mention all of these climate changes were due to natural causes.

This Chapter proves that history can be much more interesting than Grecian urns and painted ceilings, and that footprints and echoes left by our ancestors provide fascinating clues. When it comes to Archaeology, we have merely scratched the surface; so keep practicing with your bull whip, because we will explore more of history's mysteries in Chapter Eight. But before your brain overheats, it's time for some . . .

CHAPTER FIVE:
RECREATIONAL ACTIVITIES

Our ancestors realized that now and then they had to stop piling up pyramids and moving massive moai for a little recreational activity. The Mayas appreciated the significance of competitive ball games, even though some losers let it go to their heads. We need to periodically take breaks to re-create healthy attitudes, which may be why such activity is called recreation. This Chapter is designed for all of you sports nuts, who can recall scores and statistics better than dates of other events (e.g., anniversaries). To determine whether you qualify, simply ask yourself the following questions:

- When watching sporting events do you sometimes engage in jeering, yelling, cheering, screaming, crying, or destruction of innocent household objects?

- Do those pesky family conversations sometimes interfere with your ability to concentrate on Games of the Week or tournaments?

- Are the vehicles that you purchase determined by whether or not they are available in the colors of your favorite team(s)?

If you answered yes to any of the above questions, this Chapter is for you! Since you are obviously already an expert on traditional sports, we will discuss lesser-known ones you have always wanted to play. And for you jocks who are not so good with numbers, they are even listed in alphabetical order.

Exercise Guide: This "cool down" Chapter offers stress relief, but it also includes some subtle mental exercising. Several of its sections are intentionally designed to be a little more challenging, and you might want to pause briefly after reading them. This Chapter involves more curiosity than courage and should be both entertaining and informative, so have fun!

Bog Snorkeling

This exciting sport began in 1976 as the result of conversations among bar room regulars. Using only swim fins and snorkels, competitors swim two consecutive lengths of a narrow, water-filled, sixty-yard long trench cut through a peat bog. Bog snorkelers can't use any sort of swimming stroke, but have to rely entirely upon kicking with their swim fins for propulsion. Competitors who finish with the fastest times win, but bog snorkeling is essentially all about good, ~~clean~~ fun.

Events are held in Australia, Ireland, and Wales; and include the annual *World Bog Snorkeling Championship* that takes place each August in the dense Waen Rhydd peat bog. We'll forego any more details here, because we know you are anxious to get those entry forms into the mail!

Bossaball

Although this fairly new sport was invented in Belgium, there are also leagues in: the Netherlands, Romania, Brazil, Norway, Turkey, United Arab Emirates, Egypt, Saudi Arabia, Bahrain, Qatar, Chile, Poland, Germany, and Portugal. Bossaball combines volleyball, soccer, and gymnastics; and is played on strange inflatable courts equipped with two circular trampolines separated by a center net.

Teams consist of three, four, or five players each. The offensive team includes an "Attacker," who starts to bounce on one of the trampolines while the other players are on the surrounding inflatable part of the court. The offensive team members hit the ball back and forth to each other (a maximum of eight times), while the Attacker bounces as high as he or she can on the trampoline.

At some point the ball is passed to the Attacker, who then hits or kicks it into the defending team's court. If the defending team cannot keep the ball airborne, the offense scores. If the defending team does keep it in the air, they hit the ball around while positioning themselves in an offensive configuration before their Attacker tries to score in a similar manner. Teams score one point for any ball hitting the inflatable part of the opposition's court and three points if it hits their trampoline. The first team to score twenty-five points wins a game, and there are three games to a match. Now you know what to do with the kids' old trampoline!

Buzkashi

This perennial crowd-pleaser is a traditional Central Asian sport in which all the players ride horses. The object is to grab the headless carcass of a goat or calf and toss it into a scoring area while on horseback. Except for tripping a horse, there are very few rules. Players use about any means necessary to prevent opponents from scoring, including whipping them and/or their horses.

But before you play Buzkashi at your next family reunion, you should know that there are two variations of this popular sport. The easier is *Tudabarai*, because all that the winning player has to do is ride off with the carcass until he is clear of other riders. But for those of you seeking more challenges, try *Qarajai*, where the winner has to carry the carcass around a marker then toss it into a scoring circle at the other end of the field. So bring your horse, but it may be wise to leave the family's pet goat at home!

Capoeira

This may not be very common in your neighborhood, unless your neighborhood happens to be in Brazil. Capoeira dates back to the 1500s, when slaves used it to practice a fighting style without being detected, by making it look like a dance or a game. The accompanying music and fluid movements can make it almost hypnotic to watch, but it is a serious sport and martial art.

Capoeira actually began as a survival skill, designed to help escaped, unarmed slaves stay alive in the Brazilian wilderness and defend themselves against the armed colonial agents who hunted them. Runaway slaves established primitive settlements in remote and hard to find places where they continued to practice their strangely moving, yet formidable fighting technique. After slavery ended, a few practitioners became bodyguards, hit men, or joined bands of roaming thugs; but more respectable folks also kept it alive as a recreational activity and sport.

Capoeira events often begin when competitors and musicians form a circle. As they sing and clap their hands, two competitors enter the circle and start rocking back and forth to the music. Constant movements which may includes rolls and cartwheels not only make competitors moving targets, but are also meant to fake opponents into leaving themselves open to an attack. The latter may include a: direct kick to the face, takedown, elbow strike, leg sweep, punch, or head butt. Defensive moves are more likely to be those of avoidance than blocking ones. A match continues until one competitor leaves or gets replaced by another, or it is called off by one of the musicians.

When played as a game, Capoeira usually emphasizes skill, instead of knocking down or injuring opponents. If a takedown does occur, victims are often allowed to get up and continue in the game, plus kicks and punches are "pulled" before making contact. So if you are a break-dancing street musician who is also into martial arts, this could be the ideal sport for you!

* **Pause at this point to rest up after the first four sports.**

Cheese Rolling

Each year, fearless English sportspeople and equally fearless spectators gather atop Cooper's Hill to participate in the ever-popular sport of cheese rolling. This involves rolling a wheel of hard *Double Gloucester Cheese* down a hill which is so steep that the aforementioned cheese can reach a speed of seventy miles per hour. The course is so incredibly treacherous they limit the number of competitors to twenty idiots, and have emergency medical personnel ready to whisk the injured (and the cheese) off to the nearest pub!

Since they are rather complex, you might want to write down the following rules:

- Chase the rolling cheese down the steep hillside, and

- Be the first to cross the finish line at the bottom.

In 2005, races were temporarily delayed to permit ambulances to return from the hospital, and during another recent competition, six people fainted just watching the event, and four other spectators suffered minor injuries, including one who fell out of a tree and had to be taken down the slope on a backboard. Cheese Rolling is believed to have begun as a heathen festival to celebrate the return of spring, but these days it is much more civilized. The winners get cheeses, and increases in their insurance premiums.

Chess Boxing

According to the *World Chess Boxing Organization (WCBO)*, this combines the #1 thinking sport and the #1 fighting sport into a hybrid that demands the most (both mentally and physically) from its competitors. When most people think of boxers, they picture pug-nosed pugilists whose intelligent conversations consist of "Yo!" Conversely, the image most folks have of chess players are nerdy geeks wearing bow ties, ugly suspenders, and pants that are three inches too short.

Chess Boxing events, however, consist of six four-minute rounds of "speed chess," interspersed with five three-minute rounds of boxing, with each of the latter followed by a one-minute pause for the contestants to change gear (e.g., off with the gloves, on with the bow ties). The winner and losers may be decided by: a checkmate (during a chess round), a knockout (during a boxing round), exceeding the twelve-minute cumulative time limit on the chess timer, or the referee's scoring of the boxing rounds. If the chess game ends in a stalemate, the contestant with the higher boxing score wins, and if that score is equal the player with the black pieces wins. But isn't that what you would expect?

Extreme Ironing

Not to be outdone by Chess Boxing, this activity combines extreme sports with performance art. Extreme Ironing was invented in Leicester, England, of course, by a Mr. Phil Shaw who, after an especially strenuous day at the local knitwear factory, elected to do his ironing while engaged in his rock-climbing hobby. Two years later, while traveling under his nickname of "Steam," Mr. Phil promoted the activity during an international tour to the U.S., South Africa, Fiji, Australia, and New Zealand. And as most of you already know, it was in the latter that a chance encounter with a group of German tourists led to the formation of Extreme Ironing International and the German Extreme Ironing Section, or GEIS.

By now I know those competitive fluids are coursing through your veins, and you are eager, nay perhaps anxious, to experience first hand the endorphin rush only Extreme Ironing can provide. For the good of all life on this planet, however, please valiantly attempt to control your emotions while I explain more about it.

Individual or groups of participants take their ironing boards and iron articles of clothing in remote or dangerous places. These places have included: in forests, on mountainsides, while free diving, atop large bronze statues, in the middle of streets and highways, beneath the ice of frozen lakes, in a canoe, and while parachuting. Spin-offs include *Bungee Ironing* and *Extreme Cello Playing*.

✳ **Pause at this point to put away your ironing boards.**

Ga-ga

Before it was a Lady, Ga-ga was a sport that probably originated in Israel. It can be played by teams, or by two or more individuals, and is a form of dodge ball with some key differences. It may be played in any confined space, but typically takes place in an octagonal enclosure known as a "Ga-ga pit." To begin a game players bounce the ball, shouting the word "Ga" after the first bounce and "Ga-ga" after the second. After the third bounce, they rush in to make the first kill.

Ga-ga players aren't allowed to catch or pick up the ball, but must hit it with open hands. If they kick it or hit it a second time before it has struck the wall or another player, they are out. And if players hit the ball out of the Ga-ga pit without it touching the ground, sides, or another player; they are also out.

Players also get eliminated if they get struck by the ball anywhere between their knees and the ground, and there are rules to prevent them from protecting this area by kneeling down. In some variations, players aren't allowed to remain in the same position for over three seconds; and in others, kneeling players can be eliminated if the ball strikes them anywhere.

Ga-ga is a very physical sport which often results in such injuries as: blunt force trauma (from running into other players or into the walls), sprained fingers, head collisions (because players are usually looking down), and the dreaded "Ga-ga knuckle," which comes from scraping the ground or walls when attempting to hit the ball. There is a softer version called "Gooey Ga-ga," however, which involves playing the game in the mud.

In 1994 the U.S. Southwestern Regional Championship tournament was held in Texas, and in 2000 the first Ga-ga European Championship tournament was held in Portugal. Over thirty-six countries currently participate in this sport, which is officially sanctioned by the *National Ga-ga Association* (NGA).

✱ **Pause to consider who comes up with these sports.**

Human vs. Horse Races

The first of these sporting events took place in 1980, after a Welsh pub owner got into an argument about whether a runner could beat a rider on horseback if the distance were long enough. Although it is sometimes called a Marathon, the twenty-two plus mile length is not quite long enough to technically qualify.

Humans and horses each have advantages in certain parts of the rugged course outside the town of Llanwrytd Wells, Wales. Runners tend to be a bit faster when climbing and when negotiating the marshes or water, but horses have the upper hoof on flat terrain and always win in confrontations on narrow trails. Every June, hundreds of humans and up to fifty horses compete.

On race days, the runners leave the starting line at eleven a.m., followed by the horses fifteen minutes later. This is not intended as a handicap, but rather done because event organizers felt that bumping into hundreds of jostling humans might frighten the animals. In the thirty plus years these events have been held, runners have only won twice.

Kabaddi

Also known as "the game of the masses" because of its simplicity and popularity, variants of Kabaddi (with slightly different rules and names) actually date back to prehistoric times. This sport requires: speed, agility, strength, nerve, yoga skills, and above average lung capacity; but no special equipment (except for perhaps a whistle). It is played upon rectangular courts, both outdoors and indoors, and between teams that start each game with seven to eleven players each.

As defending team members link arms, the offense sends one player, called a "Raider," into the opposition's side of the court. Raiders score points by touching, unlinking, wrestling, or confining defending team members and returning to their own side of the court *while holding their breath*. Such defenders are declared "out," unless they catch the Raider before then, and remain so until their team scores a point when it is on offense.

Kabaddi was invented to help children in India fend off attacks of a respiratory ailment called croup. These days, Kabaddi is also played in: Japan, Pakistan, China, Nepal, Sri-Lanka, Bangladesh, Thailand, and other developing nations.

✱ **Pause at this point to review and to take a breath.**

Korfball

This sport was invented in 1902 by a schoolteacher from Amsterdam and is now played in fifty-seven countries. It might at first appear to be similar to basketball, in that Korfball is a team sport played on a court with a basket (called *a korf*) at each end. The object of both is to get the ball into these baskets or korfs, but that's where the similarities end. For example:

- A Korfball team is made up of four men and four women,

- The ball is not to be bounced, and players with the ball cannot run or walk,

- Players with the ball are allowed to move one foot as long as the other foot stays in the same place,

- When throwing the ball, players must not have an opponent of the same sex between himself/herself and the korf, and

- Korfball players aren't allowed to defend against players of the opposite sex.

Courts are divided into halves called "zones," and the korfs are atop poles two-thirds of the distance between the center line and the end of each zone. After every two goals the teams switch zones, the former attackers become defenders, and vice versa. Korfball matches consist of two thirty-minute periods with a ten-minute break between them.

Lawn Mower Racing

This sport was conceived in *The Cricketer Arms Pub* (where else?) by a group of young Englishmen who apparently were not very interested in Cheese Rolling. It emigrated to Australia, the U.S., and elsewhere; and by 2012, the *United States Lawn Mower Racing Association (USLMRA)* had cut swaths across the world of motorsports for twenty years.

USLMRA drivers must be at least eighteen years old, but youngsters as young as eight may compete with parental permission. All mowers must have been originally designed and sold to cut lawns, but the blades must be completely removed. Newcomers are welcome to seek technical and racing tips from the USLMRA founder, who of course is known as "Mr. Mow-It-All."

Competitors race for points, not cash, so you are not going to become wealthy in this cutting-edge sport. You may have the fastest lawn mower on your block, but before you paint racing stripes on it you ought to know that in 2010 at Bonneville Salt Flats, Bobby Cleveland set a new world speed record of 96.5 miles an hour. And after doing, so he even cut a little grass, just to prove it was a lawn mower!

Nalakatuk

Most Eskimo games involve the endurance, strengths, agility, and similar skills needed for survival and were used to teach their children they had to be tough to make it on their own. Some claim they invented Nalakatuk to spot game over the horizon, but they might have come up it with just because there were no decent shows on television. All you will need for your own Nalakatuk are:

- Several walrus skins with a hand-hold rope looped through holes in their edges,

- Friendly neighborhood Eskimo pullers, and

- ~~Victims~~ Competitors.

The latter stand in the center of the walrus skins while the pullers launch them as high as thirty feet into the air. It can be challenging just to stay upright, but skilled ~~victims~~ competitors sometimes do somersaults or flips. When deciding winners, judges look at balance, height, airborne movements (e.g., dancing, running in place), overall form, and grace.

Nalakatuk is an event at the annual *World Eskimo-Indian Olympics* in Fairbanks. It has been a spring feast activity in native whaling villages for as long as anyone remembers, and at Christmastime the jumpers throw candy and other goodies to children. However, it tends to be more popular among humans than walruses.

✳ **Pause briefly to patch the ceiling of your igloo.**

Noodling

This shouldn't be confused with the Italian pasta-making sport of Noodleing, and it involves a completely different skill set than, say, skydiving. All you need to go noodling is:

- A smelly river, lake, bayou, or swamp full of equally smelly fish;

- An IQ with the proper number of digits, and

- A blood alcohol level high enough to convince yourself that this is really something you want to do.

Noodling involves catching fish using only one's bare hands, and it is the most popular in the southern U.S. Some other names for the sport include: Grabbling, Hogging, Dogging,

Graveling, Tickling, Stumping, and Gurgling; so you must have heard of it! Flathead Catfish (that can grow up to forty pounds or more) are the most common noodling prey, because they tend to live in underwater holes.

Noodlers free-dive to depths up to twenty feet and shove their hands into catfish holes. Theoretically, the frightened catfish swim forward, trying to escape; and in doing so, latch onto the noodler's hand. If a fish is large enough, a noodler might even be able to hook his/her hand into its gills.

Noodling is frequently done in pairs, with a spotter to: help the noodler bring the fish to shore or a boat, make it a more social activity, and provide safety backup. Noodlers and their spotters often form long-term partnerships, and some of the hazards of the sport include: superficial cuts, the loss of fingers from bites and infections, drowning, and other aquatic life, such as alligators, snapping turtles, snakes, beavers, and muskrats, which take over abandoned catfish holes.

In 2011, the *Okie Noodling Tournament* was sponsored by Bob's Pig Shop and was held in Wacker Park in Pauls Valley, Oklahoma. But you don't have to go that far, because forms of noodling are legal in at least twelve states including: Alabama, Arkansas, Georgia, Illinois, Kentucky, Mississippi, North Carolina, Oklahoma, South Carolina, Tennessee, Texas, and Wisconsin. So noodle on!

Outhouse Racing

Our hardy ancestors didn't *flush* their outhouses, they *moved* them. Sometimes poor dietary decisions and gastrointestinal crises meant that this had to be done more quickly, and this naturally led to outhouse racing. Many such events, like Anchorage's *Fur Rondy Outhouse Race*, the ones during the Winter Carnival in St. George, New York, and the races held on the snow-covered Main Street of Conconully, Washington for over a quarter-century take place during the winter months.

The official rules for these may vary slightly, but in general require that the racing outhouses:

- Be of a certain minimum weight and height,

- Be mounted on runners or skis,

- Have at least three sides and a roof,

- Contain a real toilet, roll of toilet paper on a hanger, and one team member; and

- Be propelled solely by two to four others who must push and/or pull their outhouse the length of the race course.

One of the most famous winter events takes place annually in Mackinaw City, Michigan. It's billed as, "The Best Case of the Runs You'll Ever Have," and you'll see fans there wearing buttons that read, "Don't Wipe Out!" The seriousness of this sport is evident in entries with such names as the: *Wee Wee Tee Pee*, *Royal Throne*, *Shot and Squat Saloon*, a pink winged port-a-potty named the *Flamingo Flusher*, and a tropical-theme privy called the *Leaki Tiki*.

Other places like: Des Moines, Iowa; Martinsburg, Pennsylvania; Mountain View, Arkansas; and Virginia City, Nevada; hold outhouse races during the hot summer months, which lend a completely different air to the competition. Regardless of the time and place, the traditional prizes are gold-, silver-, and bronze-painted toilet seats.

✱ **Pause and review, but hold your breath if the wind changes.**

Pesäpallo

Bat-and-ball sports like Trap Ball, Rounders, Stoolball, and Cricket had been played for a long time when the first rules for American Baseball were published in 1845. Then in the 1920s, a fun-loving Finlander named Lauri "Tahko" Pihkala invented Pesäpallo (or Pesis), and things have never been the same. Today, it is Finland's national sport, and there are World Cup matches with other countries including: Sweden, Germany, Switzerland, Australia, and Canada.

If you want to get up a friendly neighborhood Pesäpallo game, you are going to need:

- Pesäpalloliitto (*Finnish Baseball Union*)-approved helmets,

- Pesäpalloliitto-approved bats,

- Pesäpallo fielder gloves, and

- Yellow balls that satisfy the specified weight and resiliency criteria and bear a Pesäpalloliitto stamp.

Teams take turns playing defense (fielding) and offense (batting). The defensive team has nine players in the field: a pitcher (at home base), a catcher (in the infield on the side of second base), three basemen, two shortstops (near second and third), and two outfielders; but all these players can switch places and their positioning varies, depending upon the batter or type of hit the defensive team expects.

An offensive team has nine ordinary batters who must bat in a pre-designated order, plus three additional "joker" (wild card) batters who are allowed to breach that order. Players often have specialized roles in the batting order, based on their abilities. Fast runners are usually placed early in the batting order, followed by players who are good at advancing runners, then those who specialize in driving runners home. Batters numbered six through nine often form another attacking combination, while "jokers" tend to be either good runners or good hitters.

Both teams have a player-Captain who tries to beat the opposing team's Captain in the "draw," which determines the team that gets to choose whether it will start as the defense or offense. Each team also has a non-playing "manager," who calls offensive plays using a multi-colored fan and directs defensive moves with hand signals. Now that you have your equipment and line-up, it's time for me to explain the simple rules of Pesäpallo.

But before I do, I probably should cover a few minor differences between it and baseball. For example, Pesäpallo is not played on a diamond, but instead on a polygon-shaped field where first base is located about where third base would be in baseball, second is roughly where first base would be, and third is more or less on the same line as first but deeper into left field. Now that you are clear on this zigzagging, I shall continue with a few more barely noticeable differences.

There isn't any pitcher's mound in Pesäpallo because the pitchers stand on the opposite side of a circular home plate from the batters, and toss the ball vertically into the air at least a meter above the batters' heads. Batters swing at the ball as it descends; and if they don't like the ball they hit on their first or second strike, they don't have to run and can continue batting. Are you keeping notes?

As in baseball, the batting team tries to score by hitting the ball and running the bases, while the fielding team tries to catch the ball and put runners out, but in Pesäpallo: a legal pitch that lands upon home plate without being hit is a strike, catching a hit ball isn't an "out" but requires that all runners return to their bases, when there are no runners on base only one bad pitch (e.g., too low, doesn't land on home plate) is considered a "walk" but otherwise it takes two, and if a batter reaches third base on his own hit he or she scores a "home run" but can remain there as a regular runner and try to score again by reaching home plate.

I could continue, but explaining all these "minor" differences is giving me carpal tunnel symptoms. So in closing, I'll simply note that as of 2009 there were over 15,000 Pesäpallo players and a lot more very confused spectators in Finland.

✳ **Take a break to review the rules for playing *Pesäpallo*.**

Sepak Takraw

This sport dates back to the 1400s, and its name comes from the Malaysian word for "kick" and the Thai word for a "woven rattan ball." Also known as Sepak Raga, Chin Lone (in Myanmar), and Sipa (in the Philippines). Similar sports include: Foot Volley, Football Tennis, and Bossaball.

It is commonly believed at least some variations were derived from an ancient Chinese military exercise, in which pairs of soldiers tried to keep a feathered shuttlecock in the air by kicking it back and both. As it evolved, the shuttlecock was replaced by a five-inch diameter ball woven out of rattan or bamboo, but these days the balls are made of synthetic material.

Early versions were not competitions, but rather were displays of skills designed to improve dexterity, exercise the body, and loosen up limbs after long periods of inactivity. It became fiercely competitive almost 200 years ago, and within a few years became part of physical education in schools throughout Southeast Asia. Sepak Takraw since spread as far as Japan, Canada, and the U.S. International play is governed by the *International Sepak Takraw Federation* (ISTAF), and the *King's Cup World Championships* are held annually in Thailand.

Modern Sepak Takraw combines hacky-sack and volleyball and is played upon twenty by forty-four foot hard-surface courts, divided by nets that are about five feet high. Teams consist of three players, two of whom play in the front row closer to the net, while the third is further away in the back row. The rules of play are similar to volleyball except players cannot use their hands, only their feet.

If the defending team fails to return a serve, the serving team wins a point. But if the serving team can't return a volley, the defending team wins the serve. Faults include: creating distractions or shouting at an opponent, jumping off the ground to serve, missing the ball on a serve, a ball that does not make it across the net, balls that clear the net but fall outside the court, playing a ball three times in a row, and players touching the ball with any part of their body except their feet.

Twenty-one points can win a set, but there must be at least a two point difference in score. Games include two sets, with a two-minute break in between, and there are three games to a match.

Shin Kicking

This sport evolved from the *Cotswold Olimpick Games,* which began in the early 1600s, and has included English family activities like: Sledgehammer Throwing, Cudgel Fighting, Coursing With Hounds, Piano Smashing, and something called Dwile Flonking. Competitors were summoned to the natural amphitheatre on Dover Hill, and cannons were fired to start the events.

The contests were refereed by officials called "sticklers," because they carried sticks to safely separate battling competitors. This is where the expression "a stickler for the rules" originated, but let's get back to the topic of Shin Kicking.

The object is to kick opponents in the shins so hard they fall down, which scores the kicker a point. The individual having the most points at the end of each timed round wins that round, and the winner of two out of three rounds wins the match. If this sounds too brutal, take comfort in knowing that contestants are permitted to pad their shins with hay. And as the timeless motto of the *Shin Kicking Association of Britain (SKAB)* says, "If it ain't broke, yer not kickin' hard enough!"

Toe Wrestling

On a particularly slow night back in 1976, patrons at *Ye Olde Royal Oak Inn* in Derbyshire inexplicably decided to hold a Toe Wrestling competition. In view of other English creations (like Cheese Rolling, Extreme Ironing, and Shin Kicking), this leads us to wonder if there is something bloody strange in England's water.

Toe Wrestling competitors take their places upon a raised platform, commonly called the "Toedium." In a gesture of courtesy and sign of mutual respect, they remove each other's shoes and socks, sit down with their bare feet flat against their opponent's, and recite a brief starting chant like, "*One, two, three, four, I declare a toe war.*" After these time-honored rituals, competitors interlock their big toes and try to "pin" their opponent's feet for three seconds, while avoiding a similar fate.

In 1997, Toe Wrestling supporters applied for the sport to be included in the Olympic Games, but their application was surprisingly denied. Not so easily deterred by rejection, however, the World Toe Wrestling Championships are still held at Derbyshire's *Bentley Brook Inn*.

* **Pause again, because the next several sports are more serious.**

Trugo

G'day mate! According to legend, Trugo was invented back in the 1920s after an Australian railway worker named Tom Grieves noticed how far the heavy rubber buffers used to connect "red rattler" train cars would roll when struck with spike driving hammers. Soon, railway workers in the western suburbs of Melbourne began to spend meal breaks competing to knock those big buffers down the eighty-nine foot length of vacant train cars.

These days Trugo is played outdoors on greens similar to the ones used for lawn bowling. A player stands on a mat with his/her back or side toward a pair of goal stakes eighty-nine feet away (sound familiar?). Swinging large wooden mallets (between their legs for men or to one side for women), players try to thwack large, wide rubber rings called "wheels" between the goal stakes, which are spaced the same distance apart as the width of those old train cars.

After a wheel passes the goal line, another player catches it in a bag on a pole. Players swap roles after every four wheels and get twelve shots from each end. The player (or team) with the most goals at the end of the timed playing period wins.

Trugo is played throughout the area around Melbourne from August until April. Its strange name supposedly comes from the expression, "true go" used to describe a successful goal. But if you are playing the women's version, it's called *Gotru*.

Underwater Football

This sport evolved from a training exercise developed by a SCUBA instructor at the University of Manitoba in the 1960s. It involved pushing a brick around the bottom of a pool, while Underwater Football uses a negatively-buoyant ball or a toy rubber torpedo. There can be up to thirteen on each team, but only five may play at a time. Player substitutions are only allowed: during a time out or referee stoppage of the game, at half time, when a goal is made, or in case of injury.

Players wear swim fins, masks, snorkels, swimsuits, and sometimes head and ear protection; however, jewelry and long fingernails are not allowed. Matches begin with players lined up on opposite goal lines with their hands on the pool ends and the ball resting upon the bottom of the pool in the center of the playing area. At the Chief Referee's signal, teams swim to the center and try to take possession of the ball. Players can come up for air at any time, *except* for when they have the ball; the ball must *never* leave the water, and most play occurs underwater.

The ball carrier may advance the ball or pass it to a teammate, as the opposing players attempt to block or tackle the ball carrier, strip the ball away, or intercept a pass. The objective is to maneuver the ball underwater to the opposition's end of the pool and place it into the goal area (e.g. gutter). Fouls include: too many players, delay of play, surface carrying, hanging onto the side of the pool, illegal substitution, rough play, equipment removal, and fighting; while penalties include: player suspension for one minute, the loss of ball possession, player suspension for two minutes plus the loss of ball possession, and ejection from a game.

Tournament matches last a total of twenty-three minutes consisting of two ten-minute periods and a three-minute halftime break, during which the teams change ends. If the score is tied at the end of the second period, an extra ten-minute period may be played to decide the winner. The halftime shows can be truly spectacular, but unfortunately they take place entirely underwater.

Unicycle Hockey

Sports played with curved sticks date back several millennia, and there are 4,000 year-old carvings the show teams of Egyptians using them to knock dates around the desert. The origin of the word "hockey" is uncertain, but it could have evolved from the Middle French word "hoquet" for a curved shepherd's staff. In any case, today there is: Air Hockey, Beach Hockey, Deck Hockey, Floor Hockey, Field Hockey, Foot Hockey, Ice Hockey, Power Hockey, Roller Hockey, Table Hockey, and of course, Unicycle Hockey.

In discussing the origin of Unicycle Hockey, the most commonly asked questions aren't "When?" or "Where?", but rather "Why?" Was this sport created by athletes who could not afford an entire bicycle? And why would anybody choose to play anything while riding the least practical form of transportation ever invented?

Unicycle Hockey is governed by the *International Unicycle Foundation*, which as you all know, publishes rules for all unicycle sports. Players ride unicycles with a maximum wheel diameter of twenty-four inches and carry sticks like those used for roller hockey. Balls are similar in weight and bounce to dead tennis ball or those used in street hockey. The rules are similar to the ones for ice hockey, but there is little physical contact (i.e., fights) between players. Two of the most common penalties are Stick In Bike (SIB) and Stick Under Bike (SUB), both of which result in a free shot being awarded to the fouled player.

Teams consist of five players (plus substitutes). Players must have both feet on the pedals to play the ball, keep one hand upon their stick at all times, and never lift the head of their stick above waist height. Playing courts are 115 to 148 feet long and 66 to 82 feet wide, with rounded or beveled corners, and goals set away from the ends so plays can go behind them.

Substitutions are allowed at any point during a game that consists of two fifteen-minute halves, separated by a five-minute break during which the teams swap ends. If the score is tied at the end of the second half, play continues for ten more minutes. If the score is still tied, each of the five current players shoots goals from about twenty-one feet until the tie is broken.

There are at least three national Unicycle Hockey leagues in Switzerland, Germany, and the United Kingdom; as well as teams in the U.S., Canada, Australia, Hong Kong, Sweden, France, Korea, Denmark, and Singapore.

✱ **Pause here to review and go find your old unicycle.**

Wife Carrying

Even though this sport also originated in Finland, it isn't <u>nearly</u> as complicated as Pesäpallo. Some claim it began with a robber named Herkko Rosvo-Ronkainen who pillaged villages with his gang in the late 1800s, and apparently had a habit of carrying off women (and sacks of food) on his back. But others maintain that it actually stems from lonesome young Finnish males sneaking into nearby villages and carrying off females to become their wives.

Wife Carrying as a sport, however, seems to have originated around Sonkajärvi, Finland, where men carrying women regularly race in pairs over an 832 foot long course containing two dry obstacles (e.g., loose sand, fences), and a pit of water at least thirty-nine inches deep. But before you send in your entry fee and start jogging about town with the (hopefully) little woman on your back, you need to know the following rules established by the all-powerful *International Wife Carrying Competition Rules Committee* (IWCCRC):

- The woman carried may be your own wife, a neighbor's, or somebody else but she must be over seventeen years old and weigh at least 108 pounds,

- Lighter women will be burdened with a rucksack that contains additional weight to bring the total carried up to no less than 108 pounds,

- The only equipment allowed is a belt worn by the carrier and required helmet worn by the carried.

- Contestants are responsible for their own safety and if necessary, insurance,

- All contestants must pay attention to instructions given by event organizers,

- In the World Championship category, the winning couple is the one that completes the course in the shortest time,

- Special prizes will be awarded for the most entertaining couple, best costume, and strongest carrier; and

- All participants must enjoy themselves.

The most common carrying techniques include: piggyback, the fireman's carry, and the ever-popular Estonian Method; in which the woman hangs upside down with her legs around the man's shoulders while holding onto his waist.

The popularity of this sport has extended well beyond Finland, with events now held in the U.S. as well as in other parts of the world. In 2011, India's first Wife Carrying competition was held in Thiruvananthapuram, and given the catchy name of "BHAARYAASAMETHAM" by wily promoters. Attendance is expected to increase as soon as people figure out how to pronounce it. And if you do plan to enter the World Championship in Sonkajärvi, before you marry any willowy waif you should know the prize is the woman's weight (without the rucksack) in beer!

Wok Racing

Contrary to popular belief, this isn't the next totally pointless and subjective event in the Southeastern, North American, International Championship competition for the title of "Non-Magnetic Chef" on The Food Channel. It actually has absolutely nothing whatsoever to do with Oriental cuisine, but rather involves racing down world-class bobsled tracks in customized, round-bottomed, cooking pans.

World Championships in both one—and four-person woks have been held since 2003 at famous alpine facilities like Winterberg and Innsbruck. Racing woks are typically imported from China, and modifications include reinforcing the bottom with epoxy and coating edges with polyurethane foam. Four-person "wok sleds" consist of two pairs of woks coupled together. Competitors usually wear heavy protective gear similar to that worn by hockey players. To improve performance, the undersides of woks are often heated before races, and riders wear ladles on their feet to further reduce friction.

Participants include celebrities and popular musicians, but well known athletes, like three-time Olympic luge champion Georg Hackl and The Jamaican Bobsled Team have also competed. In 2007, Hackl set a one-person Wok Racing record speed of roughly fifty-seven miles per hour (mph) at Innsbruck, and a year earlier the four-person record of about sixty mph was set at the same location. Now doesn't that sound a bit more exciting than stir-frying?

Zorbing

In 1994, a pair of Kiwis decided that New Zealand needed another lesser-known sport, so they invented Zorbing. The idea may have come from watching small rodents (e.g., hamsters) run inside the plastic "exercise balls" during the 1970s or maybe the apparatus constructed by the *Dangerous Sports Club* in the 1980s, but there is an excellent chance it somehow involved consumption of copious amounts of Lion Super beer or similar beverages.

Also called Sphereing and Orbing, Zorbing involves rolling along inside a double-walled, flexible, transparent plastic ball called a "zorb." The outer sphere is about nine feet, eight inches in diameter; the inner one has a diameter of approximately five feet, seven inches; and these spheres are connected by numerous ropes. You get inside a zorb via a tunnel-like entrance, and the air space between the outer and inner spheres helps to absorb shocks from running into things (e.g., trees) and over things (e.g., spectators).

The two basic types of zorbs are: harnessed ones built for one or two riders, and unharnessed ones within which up to three individuals can move around freely. Zorbing may be done on flat terrain, but it more often takes place on gentle grassy or snow-covered hills. In some cases zorbers put a little water inside so that as they are tumbling around, they also get soaked. What fun!

Today Zorbing is done in: New Zealand, Australia, England, Wales, Scotland, Northern Ireland, Estonia, Sweden, the Czech and Slovak Republics, Japan, India, Switzerland, Thailand, Canada, and the U.S. According to the folks from

Guinness, the world record for the longest zorb ride is 1,870 feet, and the speed record is 32 mph. Zorb on, sports fans!

✱ **Take a break, because recreation stuff can be tiring.**

End Game

This Chapter was purposely placed to acquaint you with enticing options to your stationary biking or aerobic gardening, and you must admit there's nothing like a day of Bog Snorkeling or Chess Boxing Polo to prepare you for another week of Web Surfing. If you did not find anything totally irresistible in the preceding list of activities, it may be time to hang up your athletic gear and devolve into a couch potato. But before you totally veg out, let's review some . . .

CHAPTER SIX:
ALTERNATIVE REALITIES

There is lots of misinformation about energies and fuels today, but this Chapter will make you wonderfully wiser and able to make more informed decisions about how you live, what you drive, and possibly even for whom you vote. It is also one of few places where the Pros, Cons, and Outlooks for common conventional and alternative energies and fuels are discussed by a professional geophysicist who is not biased by ideology, funding, or vested interests.

NOTE—This Chapter is not meant to be a complete treatise or a reference on traditional or alternative energies and fuels, but rather to exercise your critical thinking skills and encourage you to make efforts to learn more about them.

Exercise Guide: Take your time! There is a lot of information here, with much of it condensed into table form to facilitate comparisons. Pause frequently to review and take longer breaks at the indicated spots to absorb what has been covered, especially if you aren't technically-oriented. Using this approach, you should be able to complete this Chapter in a couple of days.

As in all good exercise programs, this Chapter is more difficult than some of its predecessors. Getting maximum benefit from the following exercises requires curiosity about less familiar topics, courage to question information you might have heard, and commitment to conquer challenging material. We'll begin by taking a closer look at:

Energy

Energy is officially defined as, "the property that enables something to do work," and it can be *potential* (created by position or condition) or *kinetic* (derived from motion). Types include: acoustic, chemical, electrical, gravitational, hydro (water), mechanical, nuclear, and thermal (heat), and energy can transform from one type into another. In fossil-fuel power plants, for example, chemical energy is initially converted into thermal energy to produce steam, then into mechanical energy to drive its turbines, and finally into electrical energy using its generators.

The fundamental characteristics of energy are the same whether it is derived from conventional or alternative processes; and there is very little "free energy," because collecting, distributing, storing, and converting it all require investments of additional energy. Now let's continue by exploring some common types, like:

Hydro (Water) Energy has been used for centuries. Waterwheels converted the kinetic energy of moving water into mechanical energy, but these days it is more commonly used to generate electricity. This is done by releasing the water stored behind dams through hydroelectric power plants or by "run-of-the-river" turbines in rivers, streams, tidal basins, and major ocean currents. Hydro (water) energy currently supplies approximately 19 percent of electricity worldwide, around 60 percent in Canada, and roughly 7 percent in the U.S.

HYDRO (WATER) ENERGY	
Pros:	
	1. Unlike some proposed alternatives, Hydro Energy technology exists and does not have to be developed.
	2. Once the facilities themselves are built, hydro systems offer a clean and renewable source of energy.
	3. The use of domestic Hydro Energy reduces U.S. dependence upon foreign sources.
	4. Hydroelectric power plants have relatively low operating and maintenance costs, and can be used for Base Load electricity.[1]
	5. Hydroelectric power plants can adjust their output more rapidly than other types of facilities to match changing power demands.
	6. The same water used for hydroelectric power generation can also be used for other purposes (e.g., irrigation).
	7. Some hydroelectric generators can be run in reverse as motors to pump water back up into storage reservoirs for later use.
	8. Hydroelectric power plants are likely to be able to use at least some portion of the existing electrical power grid infrastructure.
Cons:	
	1. Hydroelectric power systems require: a sufficient and reliable source of water, a large enough vertical drop, or a tidal range of at least sixteen to twenty feet in order to function effectively.

---

[1] Base Load is the term used to describe the relatively constant, or slowly varying, minimum demand for electricity within an area.

	2. Dams can be expensive to construct, can substantially alter the affected area(s), and may present a potential flooding hazard.
Outlook:	
	Changing. Hydro Energy will continue to offer a clean and reliable electricity source; but except for Canada and Turkey, construction of new large-scale hydroelectric facilities is expected to primarily be in developing countries.

The USGS has said that emphasis in the U.S. is likely to be upon smaller-scale hydroelectric plants for individual communities and facilities, but run of the river systems offer real promise in some cases and should not be overlooked. Such smaller-scale Hydro Energy systems are already being used to provide modest amounts of electricity for: communities, commercial facilities, residences, etc.

✱ **Pause and review what you've learned about** *Hydro Energy.*

Nuclear Energy is derived from such processes as nuclear fission and fusion. Current systems involve heating a fluid (e.g., water) into steam, which is used to drive turbines that generate electricity. While France relies upon nuclear power for more than 70 percent of its electricity, incidents at Three Mile Island (in 1979) and at Chernobyl (in 1986) raised public concerns to a level that the construction of new nuclear power plants within the U.S. virtually came to a halt. Interestingly, about 10 percent of U.S. electricity today is produced with nuclear material from recycled weapons under the *Megatons for Megawatts Program.*

NUCLEAR ENERGY	
Pros:	
	1. Unlike some proposed alternatives, practical nuclear energy technologies already exist and do not have to be developed.
	2. The use of Nuclear Energy reduces U.S. dependence upon foreign sources.
	3. Most nuclear facilities generate large amounts of electricity and are reliable enough to be used for Base Load power needs.
	4. If the spent fuel is reprocessed and used, 1 pound of Uranium produces about the same energy as 3.5 million pounds of coal.
	5. Technological advances (e.g., more efficient turbines) are making existing and next generation nuclear power plants more efficient.
	6. Modern nuclear facilities are designed, built, and operated with numerous safeguards against effects of seismic events, severe weather, hostile actions, and nuclear accidents.
	7. The worldwide safety record of nuclear power plants continues to be comparatively impressive.
	8. Nuclear power plants are likely to be able to use at least some portion of the existing electrical power grid infrastructure.
Cons:	
	1. It is impossible for any facility (of any type) to be 100 percent safe from all possible natural/ man-made events, and nuclear energy can produce catastrophic and long-lasting effects.

	2. It currently takes decades for western democracies to plan, design, and build new nuclear power plants.
	3. Most fission reactors use U-235, and some estimate that current amounts may only last 30 to 60 years,[2] depending upon demand.
	4. Waste heat disposition can become problematic, requiring some nuclear power plants to temporarily reduce their operating levels.
	5. Opponents have expressed concerns about the transportation of radioactive material through populated areas.
	6. Proper storing and securing nuclear waste and facilities no longer in operation are concerns that must be adequately addressed.
Outlook:	
	Controversial. There are emotional arguments on all sides of this issue. Proponents say that nuclear energy is one of the cheapest, cleanest, most abundant, and longest—term answers to U.S. energy needs; while opponents express concerns about safety, security, and the environment. We need to get past hyperbole to the facts.

In 1987, the Government began planning and construction of its Nuclear Waste Repository in Yucca Mountain, Nevada to store its spent reactor fuel and other high-level radioactive waste. In 2009, the Obama administration terminated the project, in an action the Government Accountability Office (GAO) described as being for political, not technical or safety reasons. This leaves civilian facilities with no long-term storage option, and

---

[2]   However, breeder reactor technologies that use more common forms of Uranium and Thorium could extend this to 1,000 years or more.

means that they must continue to store nuclear on-site in pools or in "dry storage casks" made of steel and concrete.

✱ **Pause here and review what we've covered about *Nuclear Energy.***

There are no clear-cut, universally-accepted, distinctions between *conventional* and *alternative* processes. Even though many people today consider the use of Hydro (Water) and Nuclear energy to be conventional, the following three are typically viewed as alternatives:

Solar Energy is currently used to heat water, buildings, and generate electricity; and there are currently two basic types of solar systems. Some collect thermal energy used for heating and to produce steam for generating electricity; while photovoltaic systems convert sunlight into Direct Current (DC) electricity. Solar heating systems may be a bit more expensive than conventional ones, but their efficiency has significantly improved in the last two decades. Photovoltaic cells are used to power satellites, calculators, lighthouses, signs, emergency roadside telephones, boats, and remotely-located buildings; and although top commercial systems may be up to 21 percent efficient, residential versions have an average efficiency of about 15 percent.

SOLAR ENERGY	
Pros:	
	1. Unlike some other proposed alternatives, basic Solar Energy technology exists and does not have to be developed.
	2. Once the equipment itself is manufactured, solar systems provide a clean and renewable source of energy.
	3. Use of Solar Energy systems can reduce U.S. dependence upon foreign sources.

	4. Some solar systems can also store thermal energy on-site for use during periods of decreased sunlight.
	5. As its initial costs decrease, solar offers residential and commercial consumers an affordable option tor augment or replacing their conventionally-powered heating systems.
	6. Solar also offers those in remote locations with modest power requirements another option for generating electricity.
	7. The cost of photovoltaic systems is decreasing, and they can convert energy more efficiently than some biomass processes.
	8. So-called "smart grid" technologies may allow networked solar systems to contribute more to Base Load power generation.
	9. Large-scale solar power plants are likely to be able to use at least some of the existing electrical power grid infrastructure.
Cons:	
	1. Solar system output depends heavily on the sunshine amount, and energy storage/backup systems are often required during nights and/or in adverse weather conditions (e.g., snow).
	2. Solar power is still substantially more costly per kilowatt-hour than conventional and most other alternative renewable energy.
	3. Performance of solar collectors is significantly affected by Sun angle, absorption of solar energy by precipitation or atmospheric particles, shading, surface dirt, and collector temperature.

	4. Output of most current solar systems is not consistent, reliable, or economically practical enough to provide Base Load power.
	5. A 4 to 12 percent energy loss occurs in converting the DC output of photovoltaic cells into Alternating Current (AC) power.
	6. Increased efficiency levels will be needed to make large-scale solar electrical power generation economically practical.
	7. Solar Energy collectors require a relatively large physical footprint, and some people also feel that they are unsightly.
Outlook:	
	Mixed. As the costs of traditional heating methods (e.g., natural gas, oil, electric) increase, more residential and commercial users are likely to employ solar methods to replace or to augment their conventional systems.

While emerging photovoltaic technologies (e.g., thin film, power plastic) appear promising, unless/until key technological advances occur, solar photovoltaic systems will likely continue to be used more for individual and/or remote applications than for large-scale power production.

✳ **Pause now and review the above information about** ***Solar Energy.***

Wind Energy has been used for many years in rural areas to pump water and to generate modest amounts of electricity. Today, hundreds of modern windmills in so-called "wind farms" and gigantic windmills provide electricity for thousands of homes. Comparatively few U.S. homes currently use wind to generate their own electricity, but some power companies allow consumers to opt for the electricity produced by wind farms or other renewable sources.

WIND ENERGY	
Pros:	
	1. Unlike some other proposed alternatives, basic Wind Energy technology exists and doesn't have to be developed.
	2. Once the equipment itself is manufactured, wind systems offer clean and renewable source of energy.
	3. Domestic Wind Energy systems reduce U.S. dependence upon foreign sources.
	4. Compared to other renewable alternative energy sources, wind competes well with conventionally-produced power.
	5. Wind systems offer users in remote locations with modest power needs an option for generating their own electricity.
	6. Site costs for land-based wind systems may be less because they can often coexist with farming and grazing activities.
	7. So-called "smart grid" technologies may allow networked wind systems to contribute more to Base Load power generation.
	8. Larger scale wind farms are likely to be able to use at least some portion of the existing electrical power grid infrastructure.
Cons:	
	1. Because output depends on sustained winds, which are seldom constant, most wind systems aren't used for Base Load power.
	2. Wind farms located offshore are likely to present additional technical and economic challenges than land-based ones.
	3. Windmills and wind farms can present hazards to wildlife.

	4. Wind farms typically require very large physical footprints, and many people feel that they are unsightly.
Outlook:	
	Promising. If the above concerns can be adequately addressed, the U.S. Department of Energy (DOE) has estimated that as much as 20 percent of U.S. electricity could come from wind energy.

Since 2010, China has been the world leader in Wind Energy utilization, leading the U.S., Germany, Spain, and others. Recent proposals include construction of offshore wind farms in areas having consistently strong winds, and new turbine designs are more efficient. There have been major technological advances, but the future of wind power might depend upon public support, the economic and political climate, and marketing factors.

✻ **Pause here and review what you've learned about** *Wind Energy.*

Geothermal Energy is used in several ways. Because the average subsurface temperature of the ground below the frost line in the U.S. is around fifty to fifty-four degrees F and relatively constant, geothermal heat exchange can be used for both heating and cooling. Geothermal Energy in the Earth's interior may also be tapped by drilling into reservoirs of hot water or steam or by pumping water into rock layers heated by geothermal processes. This can be a very important energy source in areas of geothermal activity such as: Iceland, Italy, and in some regions of the U.S. (e.g., Nevada, California,

Idaho). Natural hot water can provide heating, and geothermal steam can also be used to generate electricity.[3]

GEOTHERMAL ENERGY	
Pros:	
	1. Unlike some proposed alternatives, basic Geothermal Energy capture technologies exist and do not have to be developed.
	2. If properly designed, built, and operated, most geothermal systems are relatively clean sources of renewable energy.
	3. Domestic Geothermal Energy systems reduce U.S. dependence upon foreign sources.
	4. Many Geothermal Energy systems are reliable and economically practical enough to be used for Base Load power generation.
	5. Once built, many geothermal power plants can fulfill virtually all their own operating energy requirements.
	6. So-called "hot rock" geothermal facilities can be built virtually anywhere that one can drill down to high enough temperatures.
	7. Geothermal power plants have a relatively small physical footprints and little effect upon the natural environment.
	8. Large geothermal power plants are likely to be able to use at least some portion of the electrical power grid infrastructure.
	9. It may be possible to repurpose depleted oil and gas wells and older geothermal, plants instead of having to drill new ones.

---

[3] The EPA has estimated that using geothermal and geo-exchange systems could save as much as 70 percent on heating/cooling costs, but their initial installation costs can be quite substantial.

Cons:	
	1. The initial construction costs for both individual and commercial Geothermal Energy systems can be rather substantial.
	2. If not designed and operated properly, some geothermal power plants may contribute to air and/or water pollution.
	3. Even though the Government classifies it as "renewable," some Geothermal Energy resources are not naturally replenished.
	4. Opponents claim that some geothermal activities have caused (relatively minor) geologic instability.
Outlook:	
	It Depends. Accessible Geothermal Energy is relatively plentiful, and some maintain that identified resources could provide as much as 10 percent of our energy. The use of geothermal heat exchange systems holds promise, but the future of large-scale geothermal power plants depends upon: the relative costs, demand, and competition from other energy alternatives.

There is roughly 50,000 times as much Geothermal Energy within 10,000 meters (about 33,000 feet) of the Earth's surface than in all other natural resources; and since 1980, the costs of operating geothermal power plants have decreased by approximately 50 percent. Geothermal is currently considered one of the most important renewable sources of energy behind Hydro, Solar, and Wind, and some predict that it will eventually produce up to one-sixth of the world's electricity.

✱ **Take a break to review the information about all types of *Energy*.**

Now that you are considerably smarter than most folks regarding Energy, we'll continue by defining:

Fuel

Fuel—A substance that can store energy in a form that is relatively stable and can be more easily transported from its production site to end users.

When end users consume a fuel, they release its stored energy, often in the form of heat to power engines, cook food, warm buildings, drive generators, etc. While seeking ways to eliminate pressure vessels and water pipes required for steam engines, inventors discovered they could burn fuel to move pistons and turbines. But considerations for transportation fuels also include: (a) how much fuel can be stored onboard, and (b) how rapidly it can be replenished. We will continue now by examining variations of conventional fuels as well as more exotic alternatives.

Fossil Fuels

In 2000, fossil fuels that formed over millions of years from the remains of plants and animals provided more than 85 percent of the energy used in the world for transportation, manufacturing, generating electricity, heating, and cooking. But deposits of petroleum, natural gas, and coal are all finite, and considered to be "nonrenewable" resources.

Petroleum. Although early internal combustion engines included some exciting but unsuccessful designs that used combustibles like gunpowder, the survivors learned that petroleum products met most of their requirements pretty darn well. Gasoline, kerosene, diesel, and other fuel oils can store substantial amounts of (chemical) energy in forms that are relatively stable and are easier to transport and to replenish than equivalent amounts of wood or coal.

Depending upon its quality, current market demand, etc., approximately 70 to 75 percent of each barrel of crude oil is typically converted into transportation fuels. The rest is used for: heating, industry, electrical power generation, lubricants, kerosene, and refinery gases, coke, asphalt and road oil, petrochemical "feed stocks" used in synthetic rubber and fibers, chemicals, drugs, detergents, plastics, and other products.

PETROLEUM	
Pros:	
	1. The use of domestic petroleum can significantly reduce U.S. dependence upon foreign sources.
	2. As fuels, many petroleum products are relatively stable, easily transported, and have comparatively high energy density.
	3. Technologies and infrastructures needed to produce, distribute, and use most petroleum products are well developed.
	4. The output of oil-fueled power plants is reliable, consistent, and economically feasible enough to provide base load electricity.
	5. Using alternative methods to synthesize certain petroleum by-products can be more expensive or economically impractical.
Cons:	
	1. The world's Petroleum resources are finite and not renewable.
	2. Many Petroleum resources are in politically unstable regions and may be used as weapons in economic warfare.

	3. The combustion of many petroleum products results in so-called greenhouse gases and other environmental pollutants.
	4. Improved access to domestic petroleum oil deposits is required to help reduce the U.S. dependence upon foreign sources.
Outlook:	
	***With Us Awhile**. Because of complex dependencies on petroleum products, it is likely that they will be with us for some time to come or until we develop truly acceptable and affordable alternatives.

The reality is that we are heavily dependent upon petroleum products, even if we stopped using them entirely for fuels. The U.S has substantial domestic deposits, but access to many of them is currently limited due to the environmental lobbying of governmental agencies. If we can get past the politics and myopic ideologies, petroleum products can help us build a temporary bridge to transition to cleaner and more renewable forms of energy and fuels.

* **Pause and review what you've learned about *Petroleum*.**

Shale Oil is the term typically used to describe organic mixtures called *kerogens*, trapped within a type of sedimentary rock called shale. The largest known shale oil reserves in the U.S. are in western Colorado, southern Wyoming, and eastern Utah; and have been estimated to hold at least 1.5 trillion barrels. This is roughly 5 times larger than the proven crude oil reserves in Saudi Arabia, and more than 1,800 times that currently in our Strategic Petroleum Reserve.

Most U.S. Shale Oil lies under land controlled by the Federal Government, due to Congressional action in the early 1900s resulting in the development of the *Naval Petroleum and*

Oil Shale Reserves Program. Earlier attempts to extract shale oil used a process called "retorting," which involved: mining, hauling, and crushing the shale before extracting and processing the kerogen into gasoline or jet fuel. Retorting was environmentally hostile, and produced in some rather odd results; including ending up with larger volumes of material than was originally mined!

SHALE OIL	
Pros:	
	1. The use of domestic deposits of Shale Oil can very significantly reduce U.S. dependence upon foreign sources.
Cons:	
	1. The economic practicality of Shale Oil extraction is dependent upon the current price of conventionally-obtained oil.
	2. More environmentally-friendly Shale Oil extraction methods (see below) haven't yet been done on a large-scale, operational basis.
	3. Opponents argue that Shale Oil exploration and production may damage pristine wilderness areas in the western U.S.
Outlook:	
	Unclear. Unless such issues are adequately addressed, large-scale use of domestic Shale Oil resources is uncertain at this point.

The In-situ Conversion Process (ICP) is one way under development to extract Shale Oil. It involves drilling holes, inserting resistance heaters, and heating the subsurface shale layer to approximately 650° to 700° F for three to four years to expel the dense oil and convert lighter compounds from liquids into gases that could then be extracted via conventional wells. To prevent the process products from escaping, surrounding groundwater would have to be frozen into a "freeze wall." Similar freeze walls have already been used to prevent water from seeping into mines, Boston's Big Dig, and our Strategic Petroleum Reserve in Louisiana.

✱ **Pause briefly and review the above information about** *Oil Shale.*

Natural Gas is another type of fossil fuel found in underground deposits that may be stand-alone, associated with oil or coal, or trapped within porous rock. It must be processed to separate its largest component (i.e., Methane) from other fuels like Propane and Butane and contaminants, and it is currently used for: heating, cooking, generating electricity, and as feedstock for manufacturing plastic, paint, fabric, fertilizer, and organic chemicals. The U.S. has very substantial deposits of natural gas, but the environmental lobbying of Government agencies has made many off-limits and has required the U.S. to import more natural gas (currently around 19 percent) from foreign sources.

Natural gas can be stored underground in: depleted reservoirs from previous gas wells, salt domes, or tanks; but is hard to transport in pipelines much farther than about 2,400 miles over land or 1,200 miles over water. Unwanted, or "stranded," gas is injected into wells or burned off in a process called "flaring," but technology is under development to convert it into synthetic gasoline, diesel, or aviation fuel.

Compressed Natural Gas (CNG) is one way to utilize natural gas as a cleaner burning fuel for transportation and other applications. CNG internal combustion engines are roughly as efficient as gasoline ones, but less than modern diesels; and although some such Bi-fuel engines can use either petroleum-based fuel or CNG, they may not have high enough compressions ratio to take full advantage of the latter's octane.

NATURAL GAS	
Pros:	
	1. Use of domestic Natural Gas deposits can significantly reduce U.S. dependence upon foreign sources.
	2. A distribution infrastructure for Natural Gas is already in place throughout much of the U.S.
	3. Gas-fired power plants are reliable and economically practical enough to provide Base Load, and some Load Following, electricity.[4]
	4. With an octane rating of up to 130, CNG has the potential to optimize efficiency of engines with high compression ratios.
	5. CNG-fueled vehicles produce about 90 percent less CO and N_2O, and 30 to 40 percent less CO_2 than gasoline-fueled ones.
	6. One Gallon of Gas Equivalent (GGE) of CNG is currently about 30 to 40 percent cheaper to produce than a gallon of gasoline.
Cons:	
	1. Like other fossil fuels, Natural Gas is a finite resource and its combustion still produces some (lesser) amount of pollutants.
	2. Natural Gas is flammable, and in substantial concentrations, can kill by decreasing the amount of available Oxygen in the air.
	3. Withdrawal of large volumes of subterranean gas can cause subsidence of the overlying ground and associated damages.

---

[4] Load Following is the term used to describe electrical energy sources that can more rapidly respond to accommodate changes in demand.

	4. To carry enough CNG to be able to travel reasonable distances, it must be compressed to 3,000 to 3,600 pounds per square inch (psi), which means that onboard CNG fuel tanks must be sturdier and more expensive.
	5. Even at 3,600 psi, CNG only contains about 1/3 as much energy as gasoline and such pressures make tanks potentially explosive.
	6. Some Shale Gas extraction methods (see fracking below) are controversial, and could present serious environmental threats.
Outlook:	
	Mixed. Improved access to domestic Natural Gas deposits will be required to provide a bridge for transition to alternative energy/fuels. The demands for Natural Gas for heating, power generation, etc. are likely to compete with its use as an alternative transportation fuel, and may potentially increase the market price of Natural Gas.

Shale gas is the fastest growing component of today's natural gas market, but it often involves hydraulic fracturing, or *fracking*, which consists of injecting large volumes water, sand, and other chemicals down wells under high pressure to crack rocks and release trapped gas, and requires tanks or ponds to hold the chemically-laden "flowback" that returns from fracked wells. Although the basic process has been used since the 1940s, it has become more controversial with the advent of horizontal drilling. Main concerns include that fracturing, leaks in the cement that is used to fill the gap between the pipe and borehole wall, or the mishandling of flowback may contaminate aquifers. While there is no conclusive evidence fracking itself has polluted aquifers, there are tales of flaming faucets.

✳ **Pause here and review what you've learned about _Natural Gas_.**

Clean Coal. Coal is one of the oldest (and traditionally one of the dirtiest) fuels. Those of us who grew up in homes heated by it might recall having to carry out seemingly endless quantities of unburned "clinkers," but furnaces back then also emitted SO_2, N_2O, CO, CO_2, particulate matter (i.e., soot), and radio nuclides. As the world supply of crude oil decreases, attention will focus on ways to produce other fuel(s) from solid coal.

Such processes aren't new, because in the 1800s Boston used gas derived from coal for its street lamps; and during World War II, Germany converted coal into liquid fuel for its tanks. These days, South Africa produces most of its diesel fuel from coal; and in the 1970s, the U.S. Government sponsored work to derive gas and liquid fuel from coal. Comparatively recent technologies that are collectively called "Clean Coal," are intended to reduce the adverse impacts of using coal as an energy source while producing forms of gas or liquid fuels.

CLEAN COAL	
Pros:	
	1. Use of substantial domestic coal deposits significantly reduces U.S. dependence upon foreign sources.
	2. Output of coal-fired power plants is reliable and economically practical enough to provide Base Load electricity.
	3. Proven technologies exist to convert coal into synthetic (diesel) oil, and cleaner-burning fuels like natural gas, Hydrogen, etc.

	4. Proponents argue that Clean Coal technologies are now being developed that will allow coal to be used to produce the above and generate electricity with lower greenhouse gas emissions.
	5. Others say that CO_2 created during such processes could be sequestered deep underground, injected into oil or gas wells, or re-used by biofuel plants during photosynthesis.
	6. Large-scale production facilities are likely to be able to use at least a portion of the existing oil/gas distribution infrastructures.
Cons:	
	1. The energy density of coal is relatively low, which means that large quantities must be mined, transported, and processed.
	2. The technologies needed to produce Clean Coal and to capture and store the CO_2 are currently quite expensive.
	3. Opponents say the CO_2 sequestered underground may work its way to the surface, or that using injected CO_2 to extract oil and gas may lead to higher concentrations of CO_2 in those fuels.
	4. Even though domestic coal deposits are large, like other fossil fuels, they are neither infinite nor renewable.
	5. Opponents also argue that large coal mining activities have a number of other deleterious effects on the natural environment.

Outlook:	
	Mixed. Coal use will continue for some time and worldwide demand might increase, because it's cheap and there's a lot of it in countries like India and China. If environmental concerns are addressed and large-scale clean coal technologies are proven to be economically practical, it could also be a viable "alternative" fuel in the U.S.

One major challenge involves *economically* converting pre-combustion coal into fuels that are less harmful to the environment. Doing this may include chemical treatments of solid coal and *underground coal gasification*, during which jets of steam, air, Oxygen, and chemicals blast coal seams to produce a fuel gas that contains less Carbon, Sulfur, Nitrogen, and other components that create pollutants or greenhouse gases.

Another major challenge involves Carbon Sequestration and Storage, or CSS. One proposed method includes: capturing the CO_2 produced during combustion, compressing it into a liquid form, and injecting it deep beneath the surface into old oil fields or saline aquifers where it would hopefully be trapped by rock to prevent it from seeping back to the surface and into the atmosphere. Another proposes using the CO_2 to help grow biomass fuel plants (see Biomass Fuel).

Interestingly, a number of energy experts say overcoming such challenges will likely involve complementary efforts of interests in the U.S. and China. The U.S. can offer technological expertise and ideas, while China has proven that it can field large-scale test beds and operational facilities much more rapidly. Such contributions may be necessary to make Clean Coal practical and affordable.

✳ **Pause here and review the above information about** *Clean Coal.*

Methane Hydrate is a form of natural gas that consists of Methane molecules trapped within a crystalline sherbet-like structure of water ice. Methane Hydrate deposits are typically found beneath ocean sediments and in Arctic permafrost layers.

METHANE HYDRATE	
Pros:	
	1. Some estimate the global inventory of Methane Hydrate may be as much as ten times that of more common forms of natural gas.
	2. One cubic yard of (frozen) Methane Hydrate can produce up to 164 cubic yards of Methane gas.
	3. Use of Methane Hydrate deposits could augment finite domestic natural gas resources.
Cons:	
	1. Unburned Methane is a "greenhouse gas" and, if inadvertently released, could significantly contribute to global warming.
	2. The combustion of Methane produces CO_2 like other fossil fuels.
	3. Some claim Methane Hydrate releases have caused landslides, and drilling into such deposits could de-stabilize the substrate.
	4. Reclamation of Methane Hydrate deposits may not be reliable or economically practical enough to be used for Base Load power.
	5. There is currently relatively limited understanding of how using this potential fuel source may affect the environment.

Outlook:	
	Attractive, but. Under current technologies, most known Methane Hydrate deposits appear unlikely to be commercially exploited as sources of energy, and Methane Hydrate also faces the same challenges as other types of fossil fuels.

Norway, China, Canada, and the U.S. are all exploring for Methane Hydrate, and the USGS has reported large deposits off the Carolina coasts. Japan, which gets roughly 20 percent of its energy from coal and is the world's largest importer of natural gas, recently announced that it has successfully extracted some of the so-called "flammable ice" from 1,000 feet under the seabed approximately fifty miles South of Japan.

✳ **Congratulations! You have reached the mid-point of this Chapter. Take an overnight break to absorb the information covered up to this point.**

Non-Fossil Fuels

Biodiesel is the term used for diesel fuel made from non-fossil organic sources instead of petroleum. It's by no means new, and the diesel demonstration engine at the 1900 Paris World Exposition ran on peanut oil. Today, the most common sources of biodiesel include: vegetable oils, rendered animal fat, and used frying oil. Processing them into fuel involves removing glycerin and other contaminants through a process called *transesterification*.

BIODIESEL	
Pros:	
	1. Modern diesel engines can run on 100 percent Biodiesel with little degradation in performance, because its energy content is similar.
	2. Use of domestically-produced Biodiesel reduces U.S. dependence upon foreign sources.
	3. According to the DOE, pure Biodiesel reduces CO emissions by more than 75 percent, and a blend of 20 percent biodiesel and 80 percent petro-diesel reduces CO_2 emissions by about 15 percent.
	4. Unlike petro-diesel, Biodiesel molecules are Oxygen-bearing and can at least partially support their own combustion.
	5. The creation of Biodiesel production facilities and distribution infrastructure is underway and growing.
Cons:	
	1. Biodiesel and Biodiesel blends are typically more expensive than petroleum-based diesel on both a local and regional basis.
	2. In low temperatures, higher-concentration blends need additives or fuel warmers to prevent them from turning into waxy solids.
	3. The combustion of Biodiesel still produces some (lesser) amount of pollutants and "greenhouse gases."

Outlook:	
	Promising. About 25 million gallons of Biodiesel were produced in 2004, and this tripled to more than 75 million gallons in 2005. Given proper incentives, an increasing number of diesel powered vehicles, and public support, Biodiesel could enjoy a viable future in the U.S.

Besides rendered animal fat, Biodiesel can be made of plant oils, including those derived from: soybeans, corn, peanuts, canola, Chinese Tallow Trees, Camelina, Jatropha, Seashore Mallow, and algae. A very substantial amount of Biodiesel is produced from soybeans, but algae offers greater potential because: it contains up to 50 percent oil by weight, its single-cell plants can double their numbers in 24 hours, and algae farms may be one way to sequester CO_2 produced by fossil fuel combustion (see Clean Coal) since the plants would use it in photosynthesis.

Most Biodiesel production involves dissolving a catalyst (e.g., sodium hydroxide) in an alcohol (e.g., methanol) for several hours, mixing in plant oils, and letting it sit for about twelve to twenty-four hours to create Biodiesel, which then must be washed with acid to neutralize the catalyst. However, several technologies are being developed that may make this process more efficient and economical; and the U.S. Department of Defense (DoD) is testing a blend of Biodiesel, Ethanol, and a biomass-derived stabilizing additive that reduces problems with the alternative fuel components.

✳ Pause here and review what we've covered about *Biodiesel*.

Ethanol is also known as ethyl alcohol or grain alcohol. These days most of it is produced by: (a) the hydration of ethylene from petroleum or other sources, or (b) fermenting and distilling grain (like moonshine) or other food stock crops.

There is research underway, however, into producing Ethanol in commercial quantities from cellulosic plants such as Switch grass[5] that will not compete as much with food crops. E85 is a blend containing 85 percent Ethanol and 15 percent gasoline and is currently available in more than thirty-six states.

ETHANOL	
Pros:	
	1. Ethanol (blends) can typically be used in many newer internal combustion engines with relatively minor modifications.
	2. Use of domestically-produced Ethanol reduces U.S. dependence upon foreign sources.
	3. Ethanol burns cleaner and cooler than gasoline and has a higher octane rating, which gives it the potential for higher performance at higher compression ratios (just ask any drag racer).
	4. DOE studies suggest using Ethanol blends can substantially decrease Carbon Monoxide and CO_2 emissions.
	5. The DOE also claims that Ethanol production is CO_2 neutral, because the CO_2 produced by the crop offset the CO_2 emitted.
Cons:	
	1. Many Flex-Fuel Vehicles (FFVs) cannot take full advantage of E85's higher octane because they must also burn gasoline.

------

[5] Switch grass is a tall (up to eight feet) perennial grass native to North America. It grows along roads, in remnant prairies, and pastures. It is commonly used for soil conservation, cover for wild game, and ornamental purposes, but more recently as a biomass crop for production of ethanol and butanol and for the biosequestration of atmospheric CO_2.

	2. Pure ethanol is not volatile enough to support starting an engine in colder temperatures, which requires blending it with gasoline.
	3. E85 contains less energy than gasoline, so it requires about 1.56 gallons of E85 to go as far as 1 gallon of gas.
	4. Growing crops for ethanol can be very energy intensive and if not chosen wisely may seriously compete with those used for food.
	5. Ethanol is a strong solvent, and items exposed to it (e.g., fuel tanks, hoses) must be clean and corrosive-resistant.
Outlook:	
	It depends. If such problems can be adequately resolved and future technological advances permit sufficient quantities of "cellulosic Ethanol" to be economically produced from plant materials (e.g., wood chips, switch grass, rice straw) that don't adversely compete with major food crops or for land and/or water usage; blends of it could prove to be a viable alternative to fossil fuels.

There are several reasons why early attempts to accept Ethanol in the U.S. didn't live up to proponents' expectations. It was often derived from principal food crops (e.g., corn), it was at least as expensive as regular unleaded gasoline (especially when publicly-funded grower incentives were included), it contained less energy, and it caused serious problems in some engines. As a result, the Government chose to force it on consumers through mandatory blending with gasoline.

Business 101—*Make something people want to buy at a price they can afford.*

Brazil took a different approach after the 1970s "oil crisis," and today is a biofuel superpower. It produces Ethanol from the crops of its sugar plantation crops in distilleries powered by cane pulp. The price of Brazilian Ethanol is relatively low, but agribusiness lobbying has resulted in tariffs to keep it out of the U.S. market.

Some claim the future might not be in Ethanol, but rather in artificially-created *hydrocarbons* which are chemically more similar to fuels that already drive trains, planes, and other vehicles. These so-called "drop-in fuels" could replace diesel and aviation fuels; and unlike Ethanol, could be put into existing pipelines and fuel tanks with few or no modifications. A number of technologies are currently under development on these alternatives.

✳ **Pause briefly and review the above information about** *Ethanol.*

Methanol is also known as methyl alcohol or wood alcohol. At the present time virtually all the Methanol in the U.S. is produced from natural gas, but it can also be created using: biogas from fermenting organic matter (e.g., sewage), the CO_2 in flue gases of fossil fuel-burning facilities, coal, etc. M85 is a blend consisting of 85 percent Methanol and 15 percent gasoline.

METHANOL	
Pros:	
	1. The use of domestically-produced Methanol can reduce U.S. dependence upon foreign sources.
	2. With an octane rating of around 100, Methanol has the potential for more efficiency than gasoline at higher compression ratios.

	3. Nearly every major electronics manufacturer has plans to release portable electronics powered by Methanol fuel cells.
Cons:	
	1. Pure Methanol is not volatile enough to support starting an engine in cold temperatures and burns with a dangerous invisible flame.
	2. Many Flex-Fuel Vehicles (FFVs) cannot take full advantage of M85's higher octane because they must also burn gasoline.
	3. Methanol contains less energy than an equivalent amount of Ethanol and only 51 percent that of gasoline.
	4. Methanol refineries are typically more expensive to build than Ethanol distilleries.
	5. Producing Methanol from natural gas results in a net increase of CO_2 and liberates Carbon that might not otherwise escape.
	6. High concentrations are corrosive to light metals like Aluminum, requiring special materials for Methanol delivery, storage, & use.
Outlook:	
	Uncertain. The challenges of resolving the above issues and a very powerful lobby for Ethanol have relegated Methanol and M85 to less visible roles, but their use remains viable even on a limited basis.

Ancient Egyptians utilized Methanol in their embalming process, but today it is used to fuel Monster Trucks, Sprint Cars, Dragsters, model airplanes, and mud racers. Some say that Methanol might replace Hydrogen as the vehicle "fuel of the future" because the latter is more difficult to transport and store than initially anticipated, and support the development of Methanol (vice Hydrogen) fuel cells.

Some have suggested that Methanol could store energy produced by solar or wind systems, while others remain skeptical Methanol will ever fill more than a market niche. But since CO_2 can be used to create Methanol, it might be one way to use some of that greenhouse gas produced by fossil fuel combustion.

* **Pause here and review what you've learned about *Methanol*.**

Chemically-Generated Electricity. in this instance refers to that produced by batteries and fuel cells. Many generations of engineers have looked into using electrical power for transportation applications, and there are certainly ones for which it holds advantages over other types of fuels. Conversely, there remain a number of transportation applications for which Chemically-Generated Electricity is not currently practical. Some vehicles use braking energy to help recharge the onboard batteries, but without such a capability batteries must be re-charged for several hours at stationary charging facilities.

CHEMICALLY-GENERATED ELECTRICITY	
Pros:	
	1. Electric vehicles require no warm-up, run very quietly, and can exhibit excellent performance up to the limit of their range.
	2. Electrically powered vehicles produce no tailpipe emissions and, even when power plant emissions are included, produce only 10 percent of the pollution of equivalent internal-combustion vehicles.

	3. Electric-only vehicles are relatively cheap to refuel and cost about two to four cents a mile to operate (at the average of ten cents per kilowatt hour).
	4. Electric-only and plug-in electric hybrid vehicles can be re-charged at times when the power generation demands/costs are less.
	5. Electric motors have fewer moving parts and are about 90 percent efficient, compared to 20 percent for internal combustion engines.
	6. Use of electrically powered vehicles where appropriate may help to reduce our dependence upon foreign energy and fossil fuels.
Cons:	
	1. Batteries are relatively expensive, slow to fully recharge, and contain toxic materials that may present environmental hazards.
	2. Most current production electric vehicles have relatively short operating ranges, typically about 50 to 150 miles.
	3. A tank of gasoline contains roughly twenty times the amount of energy as a battery of equal weight and gets lighter as it empties.
	4. The U.S. power grid's capacity and infrastructure would have to be significantly improved to support many grid-charged vehicles.
Outlook:	
	Mixed. The future of electrically powered vehicles will heavily depend upon: breakthroughs in longer lasting and less costly batteries, lower prices for such vehicles, and their impact on our electrical power grid.

Although there are markets for electrically-powered vehicles, recent failures of several companies (e.g., Fisker Automotive) could indicate that past expectations were unrealistic and this technology might not be ready for universal acceptance. Also, proponents of electrically-powered vehicles often neglect to mention that both the production of their structural, mechanical, and electrical components and roughly 50 percent of the electricity currently used to re-charge their batteries involve processes that produce greenhouse gases. To offset these "carbon footprints," electrically-powered vehicles would need to be driven a significant number of miles more than logged by their typical owners.

✻ Pause and review information on *Chemically Generated Electricity*.

Hydrogen Fuel is claimed by some proponents to be the key to future energy demands. They usually mention it is our most abundant (and lightest) element, but often fail to note that there is no accessible natural reserve of uncombined Hydrogen except in the outermost strata of the Earth's atmosphere. So unless you have a space shuttle in your garage, your initial challenge is to somehow produce enough diatomic Hydrogen to make it a practical fuel, and doing that requires another source of energy.

One method is steam Methane reformation, and the most common source of Methane today is natural gas. Another involves using electricity to break apart water into its molecular components. Other methods including coal or biomass gasification, and genetically-modified organisms have also been proposed. But no matter what method you use, the end product is an extremely light, flammable gas that takes up lots of volume, is difficult to store and transport, and wants to escape to outer space or combine with other molecules. But it is precisely this last characteristic that makes Hydrogen a potential fuel. Relatively new fuel cell technologies convert the chemical energy in diatomic Hydrogen into electricity.

HYDROGEN FUEL	
Pros:	
	1. Hydrogen gas can theoretically be produced anyplace from the decomposition of water (i.e., electrolysis) but this requires energy.
	2. Combustion of Hydrogen yields mainly water vapor and alleviates many problems currently associated with burring of fossil fuels.
	3. Hydrogen could theoretically be used to power vehicles, heat buildings, cook, generate electricity, and other applications.
	4. Hydrogen is considered to be a renewable source of energy.
Cons:	
	1. Most commercial Hydrogen production today requires fossil fuel (e.g., natural gas).
	2. Hydrogen gas is currently inefficient and expensive to produce, costly to transport, and difficult to store.[6]
	3. Hydrogen gas has low energy density, so Hydrogen fuel tanks must be two to three times larger than most fossil fuel tanks.
	4. To carry enough fuel to travel a reasonable distance, Hydrogen gas needs to be compressed to at least 5,000 psi, which would result in a potentially explosive onboard tank of flammable gas.
	5. The principal end product of using Hydrogen as a fuel is Water Vapor, which itself is an already abundant greenhouse gas.

---

[6] One way to alleviate some of these problems may be to bond Hydrogen with Nitrogen to produce ammonia, which can be more easily liquefied, transported and used. However, its combustion would then produce some Nitrous Oxide.

	6. Hydrogen fuel cells are still prohibitively expensive to use as a primary transportation fuel source, and some use the precious metal Platinum of which current supplies are far from infinite.
	7. There would be very substantial costs to create a nationwide production and distribution infrastructure and produce a viable number of affordable Hydrogen-powered vehicles.
Outlook:	
	Perhaps, someday. While Hydrogen can fuel some modified internal combustion engines, it is more often proposed as a way to power fuel cells to propel electric vehicles. However, some say that practical and affordable Hydrogen fuel cell vehicles are unlikely to be widely available until around 2020.

A practical and affordable Hydrogen gas infrastructure has turned out to be more difficult than expected, leading the DOE to reduce funding for Hydrogen and fuel cell research in 2009. Many energy experts believe that Hydrogen is currently impractical as an alternative to fossil fuels.

Proponents of such vehicles sometimes neglect to mention that the production of their structural, mechanical, and electrical components usually involve processes that use energy sources (e.g., fossil fuels) which produce additional greenhouse gases. To offset these carbon footprints, such vehicles would have to be driven considerably more miles than those currently logged by their typical owners.

✳ **Pause here and review what you've learned about *Hydrogen Fuel*.**

Non-Fossil Biogas, also called swamp or marsh gas, typically consists of about 50 to 80 percent Methane, 20 to 50 percent CO_2, and other components such as Hydrogen, CO, Nitrogen, Hydrogen Sulfide, etc. It is produced by the anaerobic (i.e., without Oxygen) decomposition of organic matter, and was once collected from London sewers to fuel that city's gas street lamps. These days, if properly processed, Non-Fossil Biogas can be used for heating, lighting, cooking, and vehicle fuel.

About 7 percent of the Methane emissions in the U.S. come from livestock or poultry farms, while landfills constitute the third-largest human-related source. Manure is collected and placed into "anaerobic digesters" to optimize/stabilize Methane production, while the gas is collected by wells drilled into landfills. In 2011, there were roughly 180 Biogas recovery systems on U.S. farms; but the EPA estimates over 8,000 could do so, and thus provide approximately 1,600 megawatts of electricity while decreasing greenhouse gas emissions the equivalent of taking 6.5 million vehicles off of the road.

NON-FOSSIL BIOGAS	
Pros:	
	1. Use of domestically-produced Non-Fossil Biogas reduces U.S. dependence upon foreign resources.
	2. Non-Fossil Biogas disposes of organic waste materials to produce a gas that may be used as fuel.
	3. Non-Fossil Biogas is a renewable resource, depending upon the availability of organic waste material.
	4. Compared to many fossil and alternative fuels, Non-Fossil Biogas is relatively inexpensive.

	5. Capture of Non-Fossil Biogas can reduce the amount of Methane that is released into the atmosphere.
	6. Combustion of properly-refined and purified Non-Fossil Biogas produces less CO_2 and N_2O emissions than Diesel fuel.
	7. Producing Non-Fossil Biogas can reduce landfill waste and odors, along with the chances of water pollution.
	8. Non-Fossil Biogas may be able to use at least some portion of the existing natural gas distribution infrastructure in the U.S.
Cons:	
	1. Unburned Methane is an extremely potent greenhouse gas if it inadvertently escapes into the atmosphere.
	2. The combustion of Methane from Non-Fossil Biogas produces CO_2, another potent greenhouse gas.
	3. The production rate of Bon-Fossil Biogas isn't always predictable, and captured gas must be stored to provide a reliable fuel source.
	4. Non-Fossil Biogas must be further refined and purified to allow substituting it for natural gas as a vehicle fuel.
Outlook:	
	It Depends. A number of European countries have demonstrated the use of Non-Fossil Biogas for heating, generating electricity, and fuel for fleet vehicles. However, substantial investment may be required to make Non-Fossil Biogas a major fuel on a large scale in the U.S.

England, Germany, and Austria have all constructed Non-Fossil Biogas plants, and Sweden uses it to fuel over 5,500 (previously natural gas) vehicles. On U.S farms and landfills, Non-Fossil Biogas is mainly used for space or water heating, as boiler fuel to generate electricity, and is also sold for "carbon offset credits."

✱ **Pause now and review what we've covered about *Non-Fossil Biogas*.**

Biomass Fuel here refers to combustible solid organic materials, because we have already examined Methanol and Non-Fossil Biogas. Such solid materials include: dry garbage, plant waste (e.g., wheat chaff, corncobs), and other crops grown specifically for use as *biofuels*. Combustion of biomass has been around ever since *Zog* used a burning log to heat his cave, and wood remains the most common fuel type. Biomass Fuels are used for heating, cooking, and generating electricity; and biomass power plants are currently the second-largest users of renewable fuels in the U.S.

BIOMASS FUEL	
Pros:	
	1. Burning renewable organic waste products as fuel can alleviate the need for using non-renewable energy sources.
	2. Potential Biomass Fuel sources are available in the majority of places around our nation (and around the world).
	3. The use of domestic Biomass Fuel reduces U.S. dependence upon foreign energy resources.
	4. Burning biomass can significantly reduce the amount of space required for landfills.
	5. Biomass-fueled power plants are likely to be able to use at least a portion of the existing electrical power distribution infrastructure.

Cons:	
	1. Combustion of some Biomass Fuels can produce atmospheric pollutants that are at least as detrimental as many fossil fuels.[7]
	2. Growing large amounts of Biomass Fuel could compete for water resources and the use of agricultural land for food crops.
	3. The transportation of large quantities of biomass fuel materials to processing plants can present logistical and economic challenges.
Outlook:	
	Uncertain. It's unlikely we will ever run out of organic waste products, and using them as a fuel source seems reasonable if environmental and economic concerns are resolved.

Utilities in Hawaii have burned *bagasse*, or sugar cane waste, to make electricity for a long time; but one of the last sugar companies in the islands is considering the conversion of its plantations into "energy farms" that grow sorghum, tropical grasses, or maybe green sugar cane specifically as Biomass Fuels. A company on the Big Island has plans to harvest eucalyptus from a former sugar plantation, while another one on Oahu burns Municipal Solid Waste (MSW) to reduce the volume in that island's landfills.

Proponents claim that if properly managed, Biomass Fuel energy farms can be CO_2 neutral because growing plants absorb as much of this greenhouse gas in photosynthesis as is produced during their combustion. Progress in the use of Biomass Fuel has actually been hampered because of too

---

[7] Proponents claim that burning of biomass plant fuel is, "CO_2 neutral" because the growing plants take up an amount of atmospheric CO_2 similar to that released during their combustion.

many options, but there are still some significant challenges to be overcome.

✳ **Take a break to review what you've learned about all types of *Fuel*.**

Other Factors

A legitimate evaluation of fuels must include consideration of other factors. One way is to compare their "energy density," which describes the amount of useful energy stored per unit volume of fuel. Those with higher energy densities store more useful energy than ones with lower energy densities; and although energy densities of gases may be increased by storing them under pressure, doing so requires additional energy and introduces other issues.

But before you trade in your nifty 1978 AMC Pacer for a similarly exotic hybrid, you should know that the energy density of gasoline is roughly 6 percent better than an E10 blend, almost twice that of coal, and (for those taking notes) nearly three times that of dried camel dung. And to achieve the same energy density as gasoline you would need to compress a tank of Hydrogen gas to about 3,895 psi, so be careful when removing that filler cap! But wait, there's also the question of:

What's A GGE?

In 1994, the U.S. National Institute of Standard and Technology (NIST) defined Gasoline Gallon Equivalent (GGE) as, "The amount of (alternative) fuel it takes to equal the energy content of one liquid gallon of regular gasoline." Before then, it was difficult to compare gasoline with other fuels typically sold in different units (e.g., BTUs, KWh); but today one GGE of: electricity, CNG, Ethanol, Methanol, biodiesel, etc., all have the same energy content as a liquid gallon of gasoline.

At standard pressure and temperature, one GGE of CNG occupies roughly 948 times the volume of a gallon of gasoline, while a GGE of Hydrogen gas takes up about 2,673 times the space. To circumvent its obvious storage problems, some suggest producing Hydrogen *in situ* by the electrolysis of water, which they say takes about fifty kilowatt hours of electricity to produce one GGE of Hydrogen.

Refueling Time

This is a significant factor for vehicle fuels, because frequent or lengthy refueling times make some impractical. This section compares times it takes to replenish equivalent onboard supplies of: Fossil-Fueled, Hydrogen-Fueled, and Plug-In Electric vehicles.

Fossil-Fueled—Finding a fossil-fuel (e.g., gasoline, diesel) refueling station is relatively easy because there are so many of them. Let's suppose you roll into one just as your vehicle sucks the last vapor from your 20-gallon fossil fuel tank after driving 300 miles. On average, it will take you about 5 minutes to replenish those 20 gallons of fossil fuel, at which point you would be back on the road for another 300 miles.

Hydrogen-Fueled—Next, let's say that you have a Hydrogen Fuel Cell Electric Vehicle (FCEV). The first challenge may be locating a Hydrogen fueling station that is approved and compatible with your vehicle, and the second could be the slightly more complex filling procedures which are similar to those for CNG. Your refueling time depends on the rating of the station and capacity of your onboard tank, but once connected it should take you around ten minutes or more.

Plug-In Electrics—Lastly, suppose that you have totally drained your battery (not a good idea) on a Battery-powered Electric Vehicle (BEV). Unless you happen to be carrying a very long extension cord, your first challenge will again be locating a charging station. If you plug into a 240-volt, 40-amp circuit you might be back on the road in between 7 to 19 hours; but if you plug into a typical 120-volt, 20-amp residential circuit, you could be there for as long as 77 hours.

Infrastructure Issues

No matter how wonderful a product is, if one is unable to practically, reliably, and economically get it to its end user(s), it's relatively worthless. The risks and costs related to creating new infrastructures, modifying existing ones, or connecting to current components can be critical in determining practicality and affordability of alternatives. This section helps you appreciate how important this frequently under-emphasized factor is, and we will continue by examining some of the infrastructure issues for:

Liquid Fuels

After a western Pennsylvania well drilled by a former railroad conductor in 1859 struck oil, the latter was initially shipped via horse-drawn wagons in old whiskey barrels to nearby railway stations. By 1909, the U.S. produced more oil than all the rest of the world combined; and for much of the 19th and 20th Centuries was the largest oil producing country in the world. Today pipelines transport crude oil from oil fields and terminals to refineries where it is turned into a variety of useful products, including: gasoline, heating oil, diesel fuel, lubricants, jet fuel, and raw materials for chemicals, pharmaceuticals, and fertilizers; and pipelines transport refined petroleum products to depots, which distribute them to consumers.

Biodiesel—Links to its production and storage facilities will be needed, but the integration of Biodiesel into the petroleum infrastructure is relatively easy.

Ethanol—Blends that contain 10 to 20 percent Ethanol with gasoline could be integrated into the existing petroleum infrastructure, but higher concentrations cannot due to Ethanol's corrosiveness and tendency to absorb water.

Methanol—With changes to mitigate its more corrosive properties, the existing gasoline/diesel infrastructure could accommodate Methanol.

Gas

In 1821, William Hart drilled the first successful natural gas well below Fredonia, New York, after noticing bubbles rising to the surface in a nearby creek. It was difficult to transport natural gas very far and until the 1940s most gas discovered near oil or coal deposits was burned off, vented into the atmosphere, or left in the ground. Today it supplies more than 40 percent of the energy used by U.S. industries, and more than half of that used by commercial and residential consumers. Over 90 percent of U.S. natural gas comes from North America, and is transported through two million miles of underground pipelines.

Methane—The affordability and practicality of integrating Biogas (e.g., Methane) depends on a number of factors, including the challenges and costs associated with constructing collection and refining facilities, plus linking ones in typically rural areas to the existing gas infrastructure.

Hydrogen—The introduction of Hydrogen would require changes in virtually all aspects of the current gas infrastructure, including substantial investments in production, storage, distribution, and application facilities.

✻ Pause at this point and review *Gas* infrastructure issues.

Electricity

When the *Edison Electric Illuminating Company of New York* became our first electric utility system in 1882 its steam-driven dynamos only illuminated nearby buildings and a few private residences. Today, there are about 500 companies involved with the generation, management, and distribution of electricity over more than 186,000 miles of transmission lines. But when the generated power doesn't match the current demand, infrastructure components can shut down or fail, so the practicality of any form of energy or fuel often depends on how well the electricity produced can satisfy:

Base Load Power—plants are designed to satisfy the minimum, relatively constant demand for electricity in an area; which can change with the time of day and season. Base Load power plants are typically large conventionally-fueled (e.g., hydro, nuclear, gas or coal-fired) facilities with low(er) operating costs; but some renewable energy (e.g., geothermal, solar thermal arrays with storage, biomass) plants may also have the potential to provide some Base Load power.

Load Following Power—is associated with plants that can more rapidly increase or reduce output to accommodate changes in demand. Such plants often operate at 30 to 50 percent capacity, and might be shut down on a daily or weekly basis; while Base Load plants usually run at 70 to 90 percent capacity and rarely shut down except for maintenance. Load Following power is frequently provided by gas turbines, but some hydro, coal-fired, and newer nuclear plants may also operate in that mode.

Peaking Power—relates to systems that only run when demand is unusually high, (e.g., on a particularly hot summer afternoon) and then only for a few hours at a time. These are often smaller, faster responding, and costlier to operate units; which must be available if needed.

Smart Grid proponents claim that computerized controls will permit networked intermittent energy systems (e.g., wind, solar) to provide more Base Load power. Integration of new technologies will likely affect all aspects of the infrastructure, including power generation, transmission, operation, distribution, management, and application. This includes challenges and costs associated with establishing transmission tail segments from (possibly remote) generating sites (e.g., offshore wind farms, desert solar arrays), load balancing, and the installation of consumer interfaces to support electric vehicles, etc.

What We Know

Saving Our Society doesn't pretend to have all of the answers, but after cutting through the misinformation, agenda bias, and hyperbole about this subject we do know that:

- Fossil fuels are impossible to replace entirely in the foreseeable future. They are simply too embedded and interwoven into modern societies, and are more practical and affordable than most alternatives at this point.

- There is no single alternative solution to all our energy needs. It will take a much smarter combination of both conventional and alternative forms of energy and fuel to transition away from fossil fuels.

- An "all of the above" strategy makes a nice sounding slogan, but is unrealistic and unaffordable. The truth is that we must be a lot wiser about investing our increasingly precious resources and we cannot continue wasting them upon impractical, unaffordable, or risky technologies.

- All sides will have to be a lot more flexible to move forward. If we aren't willing to objectively examine and accept these realities, stupidity will prevail and we will all suffer the consequences.

Energy Planning

A viable energy plan that reduces our dependence upon foreign resources and stimulates our economy will require better leadership, more realistic goals, and more effective incentives; and will likely include:

- Objectively reviewing, and wherever possible, relaxing current government restrictions upon U.S.-owned companies to develop domestic resources of conventional and alternative energies and fuels that are proven, affordable, and practical;

- Terminating/suspending taxpayer-funded incentives for technologies that are no longer appropriate, counter productive, unaffordable, insufficiently mature to compete on their own merits, impractical, or too risky;

- Re-directing funds to provide *positive* (e.g., tax breaks) incentives to energy *consumers*, instead of *negative* (e.g., fines) incentives to energy *producers*.

- Providing positive incentives for U.S.-owned companies to:

  1. Practically and affordably upgrade our power grids,

  2. Convert their existing fleet vehicles or purchase ones that run on: biofuel, CNG, flex-fuel, hybrid, or battery power; and

3. Produce practical, affordable, safe, and long-lasting batteries.

- And building a near-term bridge that includes a realistic and practical combination of both conventional and alternative energies and fuels to support transition to cleaner forms that are less dependent upon foreign sources.

No energy plan should be chiseled in stone, and the above priorities are likely to change as technologies emerge, mature, and are replaced. If/when alternatives become more economically practical than conventional ones, transition to them will occur; but artificially forcing this makes about as much sense as asking an infant to run a marathon before he/ she can walk or pushing on a rope.

✳ **Congratulations! Take an overnight break to recharge your brain.**

Some claim that we need more legislation to solve our energy needs. Before you agree with them, you really should read the next Chapter, because . . .

CHAPTER SEVEN:
IT'S THE LAW!

All societies need some sort of behavioral rules. Those of primitive bands were typically based upon principles of group survival (e.g., do not eat all of our own offspring). They were often invoked by tribal leaders or shamans in the names of deities and enforced by those who wielded power over others. Since most were unwritten, they were relatively easy to modify when circumstances changed, but in the never-ending quest to make life harder these rules have turned into laws.

Exercise Guide: After all that techno-babble about energies and fuels, it seems prudent to include another "cool down" Chapter on a subject that touches us all. This one ought to be a fairly easy read requiring few, if any, pauses or breaks. However, the fascinating information it contains should stimulate your curiosity.

Ancient Egyptian law dates back to around 3000BC, and in the twenty-second century BC, King *Ur-Nammu* of Sumeria produced the first legal code consisting of "If, Then" rules. But it was Babylonian King Hammurabi whose law became most famous, and it is often viewed as the predecessor of Biblical and Jewish law. The United States has been called a nation of laws; and we are clearly one of lawyers, with more than a million licensed attorneys.

Forty percent of the current members of U.S. Congress possess law degrees; and while ignorance of the law is no excuse, apparently ignorance in creating laws is rather common. Here are some examples of what can happen with a combination of greed, ignorance, and apathy:

State Laws

Items in this section offer evidence that unless they are purchased by generous contributors, brains of state lawmakers quickly atrophy to just a few cells due to lack of use. They also prove that ludicrous legislation very easily and frequently crosses state lines.

Alabama—*Are you aware that it's against the law in Alabama to drive a car while blindfolded?* This must have been a serious problem in the past.

Alaska—*Did you know that in Haines it's illegal for a drunken bartender to serve alcohol?* It's also illegal to give alcohol to a moose, even if you're sober.

Arizona—*Do you realize that it's against the law in Arizona to hunt camels?* This is to protect any remaining survivors of the U.S. Army's Camel Corps.

Arkansas—*Are you aware that it's illegal in Arkansas to mispronounce "Arkansas"?* But it's quite legal to mispronounce, "floccinaucinihilipilification."

California—*Did you know that it's against the law in Indian Wells to drink intoxicating glue without a doctor's prescription?* This appears to be a part of California's medical mucilage movement!

Colorado—*Do you realize it's illegal to permit a dandelion to grow within the city limits of Pueblo?* And no weeds taller than ten inches will be tolerated.

Connecticut—*Are you aware that it's against the law in Connecticut to keep town records where alcoholic beverages are served?* Thish makes shensh!

Delaware—*Did you know that in Rehoboth Beach it's illegal to whisper or be rude in church?* But once you are out of that church's door, anything goes!

Florida—*Do you realize that it's against the law in Broward County for female hot dog stand attendants to wear G-strings?* Is this truly necessary?

Georgia—*Are you aware that it's illegal to bury anyone beneath a sidewalk in a Columbus City Cemetery?* I guess planning ahead isn't that important.

Hawaii—*Did you know that it's against the law to build an atomic bomb on Maui?* At first this seems like a pretty good thing, but I could be mistaken.

Idaho—*Do you realize it's illegal in Idaho to give a box of candy weighing less than fifty pounds as a romantic gift?* Was the Idaho candy-makers lobby involved with this law?

Illinois—*Are you aware that it's against the law in Joliet to mispronounce "Joliet"?* However, it's evidently legal there to mispronounce Arkansas.

Indiana—*Did you know that it's illegal in Indianapolis to ride a horse faster than ten miles per hour?* This must be why there are rarely horses in the Indy 500.

Iowa—*Do you realize that it's against the law in Iowa to charge admission to see a one-armed piano player?* So what happened to hiring the handicapped?

Kansas—*Are you aware that it's illegal to take or leave dirt at the airport in Wichita without a city permit?* They like to keep very close tabs on their topsoil.

Kentucky—*Did you know that it's against the law in Kentucky to sell dyed rabbits that are less than two months old except in quantities of at least a half dozen?* Dip me at least five more of those blue baby bunnies, Zelda!

Louisiana—*Do you realize that it's illegal in New Orleans to gargle in public?* You'll just have to drink all of those Hurricanes in your go-cups.

Maine—*Are you aware that it's against the law in Portland to tickle a woman under the chin with a feather duster?* Mainers don't need more tickled women.

Maryland—*Did you know that it's illegal to catch snot crabs in Worcester County?* You probably don't want to know any more details about this one.

Massachusetts—*Do you realize that it's against the law in Boston to take a bath unless you've been ordered to do so by a physician?* Please help me Doc!

Michigan—*Are you aware that the speed limit for ambulances in Port Huron is twenty miles per hour?* So what's with the flashing lights and siren, buddy?

Minnesota—*Did you know that it's illegal in Minnetonka for patrons to enter a massage parlor between eleven p.m. and six a.m.?* But once you are covered with oil, you may be able to just slip out the back door.

Mississippi—*Do you realize that it's against the law in Mississippi to teach another person about polygamy?* That's a danged big word fer you, Bubba!

Missouri—*Are you aware that it's illegal in St. Louis to drink beer out of a bucket while sitting on a curb?* At least find yourself a bottle and a doorway.

Montana—*Did you know that it's against the law to bring bombs, rockets, or a firearm larger than a sixty caliber to City Council meetings in Billings?* It sounds like those City Council meetings were once a lot more exciting.

Nebraska—*Do you realize that it's illegal in Nebraska to fly an airplane while drunk?* You will just have to taxi around the corn field until you sober up.

Nevada—*Are you aware that it's against the law in Nevada to use obscene language in front of a dead body?* Just cuz' you shot ol' Luke don't mean you kin cuss him out!

New Hampshire—*Did you know that it's illegal in New Hampshire to take seaweed from the beaches at night?* It's as precious to them as Wichita's dirt.

New Jersey—*Do you realize that it's against the law in Liberty Corner for a couple making out in a car to accidentally honk its horn?* Be careful out there.

New Mexico—*Are you aware that it's illegal for idiots to vote in New Mexico?* But there's apparently no law against them holding elected office.

New York—*Did you know that it's against the law in Greene to eat peanuts and walk backwards on a sidewalk during a concert?* Well, thank goodness!

North Carolina—*Do you realize that it's illegal in North Carolina to use elephants for plowing cotton fields?* At last, a law that makes good sense!

North Dakota—*Are you aware that it's against the law in North Dakota to lie down and fall asleep with your shoes on?* If you're feeling drowsy, stand up or take your shoes off.

Ohio—*Did you know that it's illegal in Oxford for a woman to disrobe while standing in front of a man's picture?* Does this sort of thing happen a lot there?

Oklahoma—*Do you realize that it's against the law for women in Schulter to gamble while naked or wearing towels?* It supposedly distracts the dealers.

Oregon—*Are you aware that it's illegal to predict the future in Yamhill?* It's also illegal there to pump your own gas, just like in New Jersey.

Pennsylvania—*Did you know that it's against the law to sleep outdoors atop a refrigerator in Pennsylvania?* This must have been a major problem.

Rhode Island—*Do you realize that it's illegal to impersonate: a Corder of Wood, a Town Sealer, or a Fence Viewer in Rhode Island?* I should hope so!

South Carolina—*Are you aware that it's against the law to sell musical instruments on Sundays in South Carolina?* You have to hum until Monday.

South Dakota—*Did you know that it's illegal in South Dakota to lie down and fall asleep in a cheese factory?* This clearly targets over-worked cheese factory workers.

Tennessee—*Do you realize it's against Tennessee law for bar owners to let unusual or obnoxious noises emanate from their premises?* Oops, pardon me!

Texas—*Are you aware that it's illegal in LeFors to drink more than three sips of beer at a time while standing?* So will you please stop sipping, or sit down.

Utah—*Did you know that it's against the law in Tremonton for a woman to have intercourse with a man while riding in an ambulance?* O-o-o-kay!

Vermont—*Do you realize that it's illegal in Vermont for married women to wear false teeth without their husband's permission?* Are those real, my dear?

Virginia—*Are you aware that it's against the law in Prince William County to park a vehicle on railroad tracks?* So what legislative genius thought this up?

Washington—*Did you know that it's illegal in Everett to display a hypnotized person in a store window?* When I count to three, step out of the store window.

West Virginia—*Do you realize that it's against the law in West Virginia to snooze on a train?* Evidently there are no tired commuters in the state.

Wisconsin—*Are you aware that it's illegal in Racine to wake up a sleeping fireman?* There will be no fires during the fire department's nap times either!

Wyoming—*Did you know that in Wyoming it's against the law for women to stand within five feet of a bar when drinking?* Excuse me ladies, but you'll have to either back up or lie down.

✱ **Pause and review the above if you have interstate travel plans.**

National Laws

To preclude American prosecutors from feeling too persecuted, I have included the following peculiar laws from other countries. If you are courageous enough to travel to any of the following places, it might be prudent to be aware that in:

Australia—*it's illegal in Australia to roam the street while wearing black clothes, felt shoes, and black shoe polish on your face?* These are considered to be tools of a cat burglar, Mate!

Cambodia—*it's against the law in Cambodia to use water guns to celebrate the New Year?* Offenders will have their water guns confiscated.

China—*it's illegal to go to college in China unless one is intelligent?* Fortunately, we are not anywhere nearly as strict in this country!

England—*it's illegal to die in the houses of Parliament?* We'll simply have none of that, old chum!

France—*it's against the law in France to name your pig Napoleon?* The French are very sensitive about their national heroes, and their pigs.

Israel—*it's against the law to raise a pig on Israeli soil?* And this law apparently also applies to any French pigs that are named Napoleon.

Italy—*it's illegal for women of "ill repute or evil looks" to enter cheese factories near Ferrara?* Elsewhere, they're evidently welcome.

Scotland—*it's against the law in Scotland to be drunk while in possession of a cow?* Perhaps it's best not to ask about this one.

Singapore—*it's illegal to sell chewing gum in Singapore?* But it's not illegal to chew it for medicinal purposes.

Swaziland—*it's illegal in Swaziland for women to wear pants?* According to the law, such pants may be ripped off and torn up by soldiers.

Switzerland—*it's against the law in Switzerland for a man to relieve himself while standing after ten p.m.?* Please have a seat, sir!

Thailand—*it's illegal in Thailand to leave your house unless you are wearing underwear?* After all, it gets pretty hot in Thailand!

Tibet—*any monk wishing to re-incarnate in Tibet must first register with the Government?* Illegal re-incarnation must be controlled!

Turkey—*it's illegal in Turkey for a man older than eighty to become a pilot?* There may be a few old turkeys, but there are no old pilots!

✱ **Pause and review the above before traveling internationally.**

Tax Laws

This section was suggested by my surgeon while he was cutting on me. Since arguing with him didn't seem to be very prudent at the time, I have included it. In 1789, Benjamin Franklin wrote, *"In this world nothing can be said to be certain, except for death and taxes."* This may be true, but since legislators create rules governing taxes, they tend to be more complicated than rules governing death. Here are a few examples:

Alabama—*Retailers who sell decks of playing cards containing no more than fifty-four cards each must pay an annual license tax of three dollars plus a fee of one dollar, and every such deck sold gets taxed an additional ten cents.*

Ancient Egypt—*The Pharaohs taxed cooking oil of peasants and also made the re-cycling of used oil illegal.*

Ancient Rome—*Under the Emperor Vespasian, operators of public toilets in Rome were required to pay taxes on urine they collected from such facilities.*

Arkansas—*Providers of tattoos, body piercings, or electrolysis services must collect 6 percent in sales taxes.*

Australia—*Prostitutes, strippers, and exotic dancers can deduct the costs of condoms, lubricants, adult toys, lingerie, and dance classes, but cannot deduct the cost of fitness classes to stay in shape.*

California—*There is a tax upon income earned by athletes, entertainers, and members of their entourages.* It was first levied on athletes from Chicago shortly after the Chicago Bulls beat the Los Angeles Lakers in 1991.

Chicago—*Fountain drinks in a glass or a cup cost you 9 percent in sales tax, but the same beverage in a can or bottle will only cost you 3 percent.*

China—*To increase revenues from cigarette sales taxes, all government officials and teachers in the Hubei Province have been ordered to buy and smoke, and there are special fines for failing to meet one's targets.*

Colorado—*As of 2010, cups and containers for food and beverage items are not taxable, but the lids for them are.*

Czarist Russia—*Peter The Great loved taxes, so he levied them on: beards, basements, bee hives, births, boots, burials, chimneys, clothing, food, drinking water, hats, marriages, and souls (my personal favorite).*

Denmark—*A tax as high as $110 per animal is levied upon owners of cows because it has been estimated that a single bovine can produce four tons of the greenhouse gas Methane per year from its burps and flatulence.*

England—*To tax people on their accumulated wealth instead of their current incomes, in 1874 the British government levied upon hats.* This led a number of milliners and hat-wearers to claim their headwear were not actually hats.

Germany—*Until recently, German tax laws allowed businesses to write off the costs of bribery on their corporate income tax returns.*

Illinois—*Besides taxes on gambling winnings, casino and track owners also pay a wagering tax which forces them to charge patrons an admission fee.*

Ireland—*To encourage artistry in the country, artists who make less than 250,000 Euros in a single year do not have to pay any taxes.*

Kansas—*Sales tax is not charged on hot air balloon rides that are piloted to some distance downwind, but it is levied upon tethered hot air balloon rides.*

Maine—*There is a tax levied upon anybody who grows, purchases, sells, handles, or processes fresh blueberries in the state.*

Maryland—*To discourage them from flushing stuff away into Chesapeake Bay, the state adds a monthly flushing tax to the sewer bills of residents.*

Minnesota—*You must pay an additional tax on the total amount received for the sale, shipping, and finance charges associated with purchase of clothing in which fur accounts for three times more of the garment than the next most valuable material.* Are you taking notes?

The Netherlands—*Citizens are permitted to deduct their costs of witchcraft training, including healing with stones, making potions, fortune telling, etc.*

New Jersey—*If you simply buy a pumpkin there is no sales tax, <u>unless</u> you plan to carve it into a jack-o-lantern.*

New York—*If you buy a whole bagel and take it home to eat it isn't taxed, but if you slice or eat it in the shop it is subject to sales tax.*

Sweden—*Until recently, the Swedish Tax Authority had the power to decide whether the names that parents chose for their children were acceptable.* And they also collect taxes from girls who earn money by stripping online.

Tennessee—*You have forty-eight hours to report and pay taxes on illegal drugs (including moonshine) you purchase, at which point you'll be provided with a stamp to affix to your illegal drugs to prove that you paid your taxes.*

Texas—*Each year before school starts some states exempt certain items from sales tax. In Texas such items include belts (but not belt buckles) and cowboy (but not climbing) boots.*

The United Kingdom—*Everyone under the age of seventy-five is required to pay an annual license fee for* *their televisions, but those who are legally blind are only required to pay half the amount.*

Utah—*Owners of businesses that employ nude or partially nude individuals to perform any service must pay a 10 percent Sales and Use Tax in addition to the usual taxes on admission and sales of merchandise, food, or beverages.*

Washington—*Until 2010, there was a tax on candy made without flour (like Rainbow Whirly Pops) while Twizzlers and Peppermint Bark were exempt.*

Closing Arguments

This Chapter illustrates what can happen when unruly children or legislators are left without proper supervision. In primitive tribes behavioral rules were frequently described by shamans, after they had entered "altered states of consciousness." These days, laws are written by politicians, who evidently do so while in similar states of mind. For those who still insist upon taking the law-making process seriously, please note the following quotes:

> "If it weren't for the lawyers we wouldn't need them."
> —William Jennings Bryan

> "If you laid all of our laws end-to-end, there would be no end."
> —Mark Twain

During your deliberations, I humbly request that you legal eagles (and beagles) consider that while it might not require much courage (or brains) to create laws like these, they are nevertheless rather curious. In my plea for leniency and as evidence of my heartfelt pledge to reform, I offer you the following Chapter.

CHAPTER EIGHT:
WORKOUT PLAN

If you have completed the exercises in the preceding Chapters, you are already smarter than most mere mortals. Your brain is bulging with a wealth of wisdom, and you are anxious to change the known world, your life, or at least your plans for the rest of the day. You now need a *Workout Plan* to keep your considerable cerebral capabilities in top shape, but where can we possibly find any follow on exercises that challenge your mental mightiness?

Suppose that you are comfortably cruising the Interstate at 65 mph in your trusty Volkswagen Bus and are rapidly catching up to a 1985 Yugo trying desperately to reach its maximum (downwind) speed of almost 52.7 mph. You glance into your rear view mirror before pulling out to pass, only to learn by doing so you would become the hood ornament on a Porsche closing at Mach Two.

Often the wisest way to move ahead involves checking where we've been, which is why this Chapter provides mental exercises about tantalizing topics that have puzzled people for many millennia.

Exercise Guide: Have fun! Pause periodically to review, and if you find some of the following topics especially intriguing, take a break and do some independent research. You might even learn the secret(s) about some of history's mysteries, like those regarding:

Ancient Giants

Many of us are familiar with the tale about the shepherd boy David and Goliath, the gigantic Philistine; but most folks are amazed to discover how common and widespread legends about ancient giants are. For example,

The Ancient Greeks called them *Titans* and *Gigantes*; the Hebrews referred to them as the *Anakim, Enim, Gog, Magog, Nephilim, and Rephaim*; the Hindus called them *Daityas,* the Navajo people referred to them as *Anaaye* or *Yeitso,* Celts called them *Fomorians,* in Wales they were referred to as the *Cawr* and *Wrnach,* the Cherokees called them *Nunhyunuwi,* the Russians referred to them as *Bogatyr, Volot,* and *Velikan;* in Spain they were called the *Famangomadan or Albadan,* the Sri Lankans referred to them as the *Gotaimbara, Mahasena,* and *Yodayo;* Paiutes called them the *Si-Te-Cah,* the Turks referred to them as *Dev Erhan,* Indonesians called them the *Buto* and *Raksasa,* the Iroquois referred to them as *Dehotgohsgayeh,* Swedes called them the *Jätar,* Bulgarians referred to them as *Ispolini,* the Basques called them *Basajaum,* in Micronesia they were called the *Puntan,* and they had other names elsewhere.

The sheer number of legends about ancient giants from so many geographically-separate cultures from around the globe is difficult to ignore, but there are other similarities also. Most ancient giants reportedly stood between seven and nine feet tall, but a few of them supposedly reached heights of twelve feet or more. They were muscularly built with several times the strength of ordinary humans, and rather aggressive. Some were said to have blonde or red hair, or strange physical characteristics like several rows of teeth or extra digits on their hands and feet. They possessed exceptional capabilities, which resulted in many of them being regarded as gods; but they could get sick, die, or sometimes even be killed. But if ancient giants actually existed:

Who Were They?

Some claim that ancient giants were mortals endowed with unique knowledge and abilities by rebellious Titans, like *Prometheus* in Greek mythology. Others say at some point our DNA was altered by extraterrestrials who visited the Earth and introduced early humans to language, agriculture, mathematics, astronomy, and such powers as levitation; while some maintain that ancient giants resulted from divine beings or "fallen angels," such as *Azazel* described in the Book of Enoch interbreeding with mortal women.

"The Nephilim were on the earth in those days, and also after that, when the sons of God came unto the daughters of men, and they bare children to them: the same were the mighty men that were of old, the men of renown." (Genesis 6:4)

A less romantic explanation says giants were products of interbreeding among certain human gene pools. If the members of one group (predominantly tall for generations) parent children with those from another group (similarly dominated by massively muscular members), it seems reasonable to expect at least some of their offspring might be a little larger than average. This could continue until other traits were introduced into their reproductive gene pool. But, if so:

Where Did They Live?

As children we were told giants lived atop beanstalks, only to later learn that they were jolly, green, and stood astride vast valleys of vegetables. The supporters of "fallen angel" theories often claim ancient giants inhabited lands near centers of early religions, while the proponents of "extraterrestrial visitor" theories typically suggest giant beings lived near various ancient (landing) sites about our planet. Unexplained feats have been attributed to them; and oversized weapons, tools, and larger than average remains have been reported in both hemispheres.

Upon returning from the land of Canaan, scouts told Moses, *"Everybody we saw there was huge. We even saw the Nephilim giants. Alongside them we felt like grasshoppers, and they looked down on us as if we were grasshoppers."* (Numbers 13: 31-33)

Ancient giants are part of a number of legends among Native American people whose ancestors lived in regions including the Great Lakes, Ohio River Valley, and the Southwest; and historical records left by our early pioneers describe finding skeletons with skulls that easily fit over an average man's head.

During the summer before my sophomore year in college, I helped a local dentist (and amateur archaeologist) excavate a burial mound on a bluff above the upper Mississippi River. After the typical sequence of probing rods, shovels, trowels, teaspoons, and toothbrushes, we unearthed human remains arranged in the fetal position. We noticed the femurs (upper leg bones) looked longer than most, and when the skeleton was laid out it measured about seven feet.

The skeleton was donated to the public high school's biology department. I don't know whether or not this particular individual would be classified as a giant, but if ancient ones did exist:

What Happened To Them?

There are obviously fewer giants around these days, except for in the NBA, and there is no shortage of hypotheses about why this is the case. Some claim that they were killed off in great battles with the smaller, yet smarter and more agile humans; while others say they were eventually assimilated into the population. Another interesting hypothesis is that because they were so big and mean, the obviously smarter humans put giants on the front lines in battles (e.g., Goliath), which resulted in fewer of them being left to propagate. A number of legends say that ancient giants became unruly and displeased the gods, who attempted to eliminate them one or more times. Apparently, one popular method that gods used for wiping out pesky giants back then were:

Great Flood(s)

The capture and release of liquid water during major Ice Ages and their warmer interglacial periods, respectively, have lowered and raised sea levels hundreds of feet, and water from melting glaciers has turned valleys and basins into ancient lakes and inland seas. But the possible cause(s) of great floods are not nearly as intriguing as the following:

Old World Legends

Most folks are familiar with the story of *The Great Flood* in the Book of Genesis, but few realize that there are very similar legends from almost every region on Earth. Here are just a few examples:

- **from Africa (Tanzania)**. Once upon a time rivers began to flood, and god told the people to get into a boat and take many animals and seeds with them. The waters eventually covered the mountains, but finally stopped rising. One of the people let loose a dove that returned. When they next released a hawk that didn't return, they left the boat and took the seeds and animals with them.

- **from Mesopotamia (Babylonia)**. In the *Epic of Gilgamesh,* which was written and compiled sometime between about 1800 and 1000BC, Gilgamesh meets an old man named *Utnapishtim*, who says the gods warned him a great flood was coming and told him to destroy his house and build a large ship. He was also told to take his wife and his family, male and female animals of all kinds, and supplies on board, and to cover the ship with pitch. It rained for six days and nights, and the ship finally came to rest atop *Mount Nisir*. After seven more days had passed, Utnapishtim released a dove, then a swallow, and finally a raven. When the latter didn't return, Utnapishtim left the ship.

- **from Ancient Greece**. *Zeus* decided to destroy humans because they had become too proud, but *Prometheus* warned his son *Deucalion* and the latter's wife *Pyrrha* and placed them into a large wooden chest, along with provisions. The rains came and lasted for nine days and nights until the entire world was flooded, except for the peaks of *Mount Olympus* (the home of the Gods) and *Mount Parnassus*. When the chest came to rest upon the latter, Deucalion and his wife got out and lived on the provisions in the chest until the water had subsided. Afterwards, Zeus instructed them to re-populate the Earth by throwing stones over their heads. The stones that were thrown by Ducalion became men, while those thrown by Pyrrha turned into women.

- **from India**. Once, a man named *Manu* saved a small fish from the jaws of a larger one, and in gratitude the fish told Manu that a great flood would soon come and destroy everything on the Earth. The fish also asked Manu to put it into a jar for protection, and each time it outgrew one jar Manu placed it into a larger one. After the fish had grown into one of the largest fish in the world, it told Manu to build a big ship because the flood was coming very soon. When the rains began, Manu tied a rope onto the fish. It guided his ship while the

Earth was covered, and after the water had receded the fish led Manu to a mountaintop.

Many biblical scholars believe that the story in Genesis was written between 550 and 450BC and was basically a re-wording of the Mesopotamian version. Young Earth Creationists place the time of the Great Flood at around 2300BC and say geological processes, including the erosion of the Grand Canyon, the uplifting of mountains, stratification of sediments, layering of fossils, etc., occurred at greatly accelerated rates as the flood waters receded. Many modern scientists respond by saying there is no proof that such geological processes were accelerated, or conclusive scientific evidence that the entire Earth was underwater at the time.

There is little question that wide-spread floods have occurred throughout history, and because so many ancient cultures lived near bodies of water, some suggest such floods were more regional than global. For example, there is archaeological evidence of inundation of an approximately 400-mile long by 100-mile wide area Northwest of the Persian Gulf in roughly 4,000 BC, however, and that could have constituted the whole world for inhabitants of the plains back then. In any case, while it may be relatively easy to envision how versions of such legends could have been shared among societies in Old World, it is much more difficult to explain the following:

New World Legends

- **from Mesoamerica (Aztecs)**. A very pious man named *Tapi* lived a long time ago. The creator told him to build a boat and to take his wife and a pair of every animal that lived onboard. Everybody thought he was crazy until the rains began. Humans and animals tried to climb the mountains, but the flood water covered them as well. After the rain ended, Tapi decided that it had dried up when a dove he released didn't return.

- **from North America (Delaware People).** At first the world lived in peace, but an evil spirit came and brought a great flood that submerged the Earth. Some people had taken refuge on the back of a large turtle so old that its shell was covered with moss. They asked a loon to dive beneath the water and bring up land, but it only found a bottomless sea. The loon flew off, but returned with a small piece of earth in its bill and guided the turtle to where the land was dry.

- **from South America (Incas)**. Over time the people became evil and got so busy doing evil deeds that they neglected the gods. Two brothers who lived in the highlands noticed their llamas acting strangely and were told that a great flood was coming which would destroy all life. The brothers took their families and flocks into a cave in the high mountains. The rains began and continued for four months, but as the waters rose the mountain also grew to keep the cave dry. Eventually the rain stopped, the mountain returned to its original height, and the shepherds emerged to re-populate the Earth. But the llamas still remembered the flood, which is why they prefer to live in the highlands.

Such stories were reportedly part of oral traditions in the Americas long before their discovery by Columbus in the decade between 1492 and 1502AD, and the striking similarities in areas like: advance warning, construction of a vessel, the stowage of animals and inclusion of family members, and the releasing of birds to determine when the flood waters had subsided suggest they might have been brought to the Americas by individuals familiar with Old World versions. But while we are talking about things getting inundated, it is a good time for us to discuss:

Atlantis

Around 360BC, the Greek philosopher Plato wrote about an ancient civilization in a place called Atlantis; but he said that the story actually originated during the sixth century BC, when an Athenian lawmaker and poet named Solon met a priest who translated it from Ancient Egyptian into Greek.

What Was It?

Plato said Atlantis was a large (345 mile by 230 mile) land mass consisting of concentric islands that were separated by moats and linked by a canal that penetrated to a mountainous central island. All of its passages were guarded by towers and gates, and stone walls covered with metals surrounded each of the rings. It was home to many exotic animals, had abundant natural resources, and was reportedly a great naval power that had conquered many parts of western Europe and Africa.

Who Lived There?

According to Plato, the gods divided the world among them, and *Poseidon* was given Atlantis. He fell in love with the mortal woman *Cleito*, and built her a palace atop its central island. She bore him five sets of male twins; the eldest of whom, *Atlas*, was made king of Atlantis and great ocean beyond (named the Atlantic in his honor). His younger brothers became princes who controlled parts of Atlas' kingdom, and together these half-human descendants of Poseidon constituted the first royal family of Atlantis.

Poseidon established laws for them and all the subsequent rulers to follow. They were to meet annually and sacrifice a sacred bull (remember the Minoans?) Its blood was blended with wine and drunk, as each ruler pledged an oath to judge according to those laws. As long as they lived and judged by Poseidon's laws, Atlantis remained prosperous and peaceful. Nearly all its needs could be met from its fields, forests, and mines; and almost anything else was imported.

What Happened To It?

Trouble started when Poseidon's laws began to be neglected. Pride overtook the rulers of Atlantis who became obsessed with power and formed an evil coalition to wage war against other nations in Europe and Asia. Plato's story claims that their failed attempt to invade Athens infuriated the gods, who one terrible night destroyed Atlantis and caused it to sink into the sea, leaving an impassable mud shoal. Some claim the decline of the Minoan culture on Crete following Thera's eruption might have been a basis for the Atlantis legend; but the location, time frame, and island's size at first glance all appear to be inconsistent with Plato's account. However, if the translation from Ancient Egyptian into Greek mistook "hundreds" for "thousands," this hypothesis becomes more feasible.

Where Was It?

According to Plato, Atlantis was located in front of the *Pillars of Heracles* (a.k.a. Hercules). There have subsequently been claims it was in such diverse places as: The Azores, northwestern Africa, the Americas, the Bahamas, the Canaries, off Cuba, in Cyprus, India, Indonesia, Ireland, Israel, Malta, Sardinia, southern Spain, Sicily, Turkey, and even Antarctica. Scientific evidence has refuted many of these, but the Atlantis legend has nevertheless persisted for over 2,300 years.

Regardless of what you believe about Atlantis, there is evidence at least some traditions and knowledge seem to have migrated between hemispheres. Could this be a sign of:

Early Visitors?

Today it is generally accepted that the Vikings reached at least as far as Nova Scotia, Newfoundland, and Labrador sometime around 1000AD, but some say the New World had much earlier visitors. Several of the most fascinating (and controversial) theories about this include those pertaining to:

Minoan Miners

Many of the ancient cultures discussed in Chapter Four flourished during regional "Bronze Ages," when stone implements, weapons, etc. were replaced by ones made of an alloy consisting of 80 to 90 percent Copper and 10 to 20 percent of another metal (often Tin). This was a major technological breakthrough, but how early metallurgists discovered the bronze-making process remains a mystery.

Ancient armies ordered thousands of bronze swords and helmets, and a Roman soldier's uniform was said to contain forty-eight pounds of the harder new metal. Demands for Copper and Tin exploded and gave rise to major trading networks. Tin ore from Cornwall, England, was shipped as far as the eastern Mediterranean and Asia, and the requirements for Copper were even greater than those for Tin.

Principal players in this lucrative metal trading business included some maritime powers like the Minoans and Phoenicians, and some insist that their enterprises extended well beyond the Old World. They claim European mines could not have kept pace with the demands and, as strange as it may sound, that ancient miners established a "Copper

Culture" in the Great Lakes region of North America where some of the purest Copper ore in the world is found and individual pieces weigh as much as 34,000 pounds.

Thousands of ancient mines have been discovered there, many of which reveal evidence of activity at least 4,500 years ago; and the methods used to extract the ore seem to be similar to those used then by miners in northern Europe. In 1894, a tablet was found near Newberry, Michigan, which some claim contains Minoan symbols; and in 1992, a harbor with remnants of a 1,640 foot pier was found on Isle Royale in Lake Superior, where a number of ancient mines are located.

Interestingly, early mining in the area evidently ceased rather abruptly at about the same time that the Bronze Age ended in the Old World, and the indigenous *Menomonie* people have a legend about mines being worked by "light-skinned men." But were these ancient strangers simply copper collectors, or:

Foreign Gods?

The Teotihuacáns, Mayas, Aztecs, and Incas all had gods that they described as having light complexions and beards; and who reportedly provided their primitive ancestors with laws, writing, knowledge of astronomy and agriculture, and other skills which enabled them to become civilized. For example,

- In his human form, the Teotihuacáns' feathered serpent god *Quetzalcoatl* had pale skin, long hair, and a beard. After losing a fight with *Tezcatlipoca* he left Mexico in a boat sailing to the East, promising that he would return again.

- *Kukulcán* was one of the Mayas creator gods. In human form, he was about six feet tall with white or silver hair, blue eyes, a light complexion, and fine features. After spending time upon the Earth he returned to the sea, but he promised to come back someday.

- When Cortés offered him gifts, an emissary from Aztec Emperor *Montezuma* asked for one of the Spaniard's helmets instead saying: "*I must show this to the Emperor, for this helmet looks like the one once worn by the white god.*"

- The Inca's light-skinned creator-god, *Virakocha,* had a beard and green eyes, dressed in a white robe, and wore the Sun like a crown. He walked the Earth disguised as a beggar teaching and performing miracles. He was persecuted by some and was said to have walked on the water. One day he went away, but his followers believed he would return.

Did other pale-skinned, bearded strangers visit the Americas many years before the Vikings, or are these simply "creative interpretations" of history by European missionaries? Before you answer, you may wish to also consider the following:

Other Evidence

- The Olmecs worked in jade and asphalt mosaics like the Minoans, traded in amber like the Phoenicians, and the helmets on their Colossal Head statues are similar to ones worn in: Sumeria, Egypt, western Africa, and Crete.

- Several ancient civilizations in the Americas carved stone faces with features (e.g., beards, noses) that do not resemble those of their indigenous people; and there have been reports of mummies found in the Americas with blond or reddish hair and beards, neither of which are Native American characteristics.

- Nicotine and Cocaine have reportedly been found in mummies from Ancient Egypt, although the plants that produce them are just native to the Americas.

- Serpents play key roles in the religions of many civilizations, including ones in the Mediterranean, northern Europe, Africa, India, Asia, and the Americas.

- Ancient rulers in the Americas wore crowns, sat upon thrones, held scepters, and were carried about on litters, just like those in the Europe and Asia.

While none of the above offer conclusive proof, collectively they seem to indicate there might have been some earlier transfer of knowledge, beliefs, and traditions. In 1997, mitochondrial DNA evidence was found in human remains dating back to before 1000BC that indicated genetic links between some early inhabitants of North America and at least two different groups of Indo-Europeans. But before we get hung up with haplogroups, let's continue our search for additional clues in more substantial things like:

Megaliths

A surprising number of ancient societies used large stones to: serve as tombs, monuments, or territorial boundary markers; support religious or astronomical activities, function as structural components, or perhaps even harness natural energies. Anthropologists have suggested some megaliths were constructed to inspire unity within societies; others claim they prove the existence of giants or aliens; and some say the builders just wanted their work to last. Perhaps the ancients constructed megaliths simply because they (somehow) *could*.

Answers to questions about why early builders chose to use multi-ton stones instead of lighter-weight materials or smaller components have eluded curious people for centuries; but these very same big blocks of rock could contain clues to some of history's other mysteries, if one knows where to look. However, we first need to separate megaliths into the following two basic categories:

- *Primitive* megaliths, which include standing or stacked stones that are more or less naturally-shaped (e.g., portal tombs, passage graves, henges), and

- *Sophisticated* megaliths, which include both massive carvings (e.g., statues, obelisks) and precisely-shaped structural components (e.g., building blocks, lintels, gables).

As you will learn, making this distinction between primitive and sophisticated forms can help reveal clues in the answers to other common questions about megaliths like:

Where Were They Built?

Differences begin to appear immediately. Primitive megaliths can be found in Europe, the British Isles, Africa, the Middle East, northern and southern Asia, India, Indonesia, Korea, Melanesia, Japan, and elsewhere. Examples include: *Stonehenge* in England, *Nan Madol* in Micronesia, the *Carnac* stones in France, *Jharkhand* in India, and others in such geographically-remote locations as Korea, Denmark, China, Russia, and North America.

Conversely, sophisticated megaliths were more common in the Mediterranean region and in South and Meso-America. Some examples of these include: the *Unfinished Obelisk* in Egypt, *Colossal Heads* carved by the Olmecs in Mexico, and structural blocks in walls, pyramids, temples, etc. in both hemispheres.

One of the world's most mysterious megalith sites lies in the Bolivian highlands near Lake Titicaca, legendary home of the Inca's creator god Virakocha. While centuries of natural catastrophes (e.g., earthquakes, floods), stone scavenging, and amateur excavations have turned it into a puzzling pile of rubble, the detail and precision of the sophisticated megaliths at *Puma Punku* are incredible. The square cuts and inside corners, evenly-spaced circular holes, and the apparently machined finishes of the stones surpass any others in the western hemisphere. We will return there later, but now let's focus on the next megalith question of:

How Were They Built?

Some primitive megaliths could have been built by ordinary humans, using brute force and simple tools. If a strong human could lift around 200 pounds, it seems reasonable that a ten foot tall, heavily-muscled giant might be able to lift roughly twice that weight. This would mean five such giants should be able to lift a one-ton stone, but even if your tribal shaman came up with really long levers and a hoist, moving more massive megaliths with such methods appears unlikely.

Sophisticated megaliths, however, require many more skills because they have to be: quarried, moved, shaped, lifted into place, and frequently fitted together (often without any mortar or cement) well enough to provide structural integrity. Colossal Heads can weigh up to 50 tons, one wall block at *Sacsayhuaman* in Peru is estimated to weigh over 120 tons, and Egypt's Unfinished Obelisk is estimated to weigh almost 1,200 tons. These are not exactly pea gravel!

Stones weighing as much as 144 tons were somehow moved to Puma Punku from quarries over six miles away, without the use of wheels or trees as rollers. In the 1500s, a native reportedly told a visitor that the place was built long before the Incas by "Sky People": beings with supernatural powers who raised stones off of the ground and floated them from the quarries "to the sound of a trumpet." This is by no means the only reference that legends make to moving rocks with acoustic energy. Ancient Greek historians claimed the walls around Thebes were built by *Amphion*, who was able to move large stones by playing his golden harp; and the Bible tells of Joshua toppling the stone walls of Jericho by having priests play trumpets and soldiers shout. There is nothing like a few levitation tunes to make moving big rock blocks easier, but let's continue our search for clues by examining the next megalith question of:

When Were They Built?

The answers to this question reveal even more differences. Primitive megaliths generally appeared earlier, and some of the oldest include the *Ġgantija* temples (supposedly built by giants) on Malta, and those at *Göbekli Tepe* in southeastern Turkey that date back as far as 10,000BC. By comparison, the *Stonehenge* is estimated to have been built sometime between about 2000 and 3000BC.

Sophisticated megalith construction generally began thousands of years later. In ancient Egypt, for example, sophisticated megaliths started to appear sometime around 2800BC; but their construction declined markedly in about 1200BC after Egypt suffered relentless attacks by naval raiders called "The Sea People." The latter rather mysteriously vanished after being defeated by Ramses III; but their attacks had severely weakened Egypt, a principal adversary of the Phoenicians, whose major cities went untouched.

This left the Phoenicians in the position to extend their influence and knowledge for centuries to come. This might not seem to be relevant, unless one considers that:

- No sophisticated megaliths in the Americas are currently estimated to date back more than about 1200BC,

- Older ones contain larger components than newer ones,

- Some of the oldest ones show signs of modern tools and workmanship, and

- Some of the highest quality megaliths are not the most recent ones. Could these be clues to help answer the next question of:

Who Built Them?

It seems that primitive and sophisticated megaliths were built by different groups, with different skills, and using very different methods As previously noted, some primitive ones could have been built with brute strength and rudimentary tools. But does this, along with their wider distribution and earlier appearance, mean they were built by the ancient giants described in legends around the globe?

Conversely, the more limited distribution, different timelines, and workmanship of sophisticated megaliths seem to indicate that the necessary knowledge and skills migrated from the Old World to South and Meso-America. But if this is the case, who were these mysterious master masons?

Carved heads, embedded in the walls of sunken temple ruins not far from Puma Punku, reflect the features of many races, and several seem to resemble aliens with large eyes and elongated skulls. In its center is a statue of Virakocha with a moustache and beard, neither of which are characteristics of indigenous people in the Andes highlands. In the Twentieth Century, a worker found a bowl with strange markings that were determined to be similar to ones in a proto-Sumerian style of writing used over 5,000 years earlier on the other side of the Earth in Mesopotamia, one of the so-called "Cradles of Civilization."

Construction of sophisticated megaliths in the New World continued under the Olmecs, Toltecs, Mayas, Aztecs, and Incas; but dropped off sharply soon after the arrival of the Conquistadors. How such incredible knowledge could be lost has puzzled people for centuries, so this is a good time to discuss:

Lessons About Losses

One of the most valuable lessons history offers is that civilization doesn't always move forward. Successful societies have succumbed to natural catastrophes or been unable to adapt to local environmental changes, and relatively advanced cultures have been conquered by more aggressive (and primitive) ones. Such conquerors often foolishly destroy irreplaceable information and alter historical records, and new regimes impose their biases and beliefs. When these occur, priceless knowledge is lost, and civilization often takes a big step backward.

Change is not always positive, and sometimes societies misplace their moral compasses. Cultures all around the globe have experienced "Dark Ages" and greed, ignorance, and apathy are evident in all cases. The preceding exercises are actually not as much about giants, floods, Atlantis, visitors, or megaliths as they are about reminding us to check our rearview mirrors and learn from where we have been. Solving them should keep your brain busy awhile; but if you run out of mental challenges, here are a few more:

Modern Mysteries

Solving the following will continue to exercise your clear-thinking capabilities. All of them are from North America, and they prove that persistent puzzles are not always ancient.

Coral Castle

Some say that the only sophisticated megalith (earthen mounds aren't stone) in North America is in South Florida. There (almost four centuries after the Aztecs and Incas, a little Latvian named Ed somehow single-handedly quarried, moved, carved, stacked, and fit together (without mortar) more than 1,100 tons of stones to build what is today known as *Coral Castle* as a tribute to his long-lost love.

He worked all by himself and at night, using tools he made out of junk. Ed wrote pamphlets about moral education and magnetic current, and spoke of something that he called a "perpetual motion holder." When asked about how he moved the huge stones, Ed would simply say, "I understand the laws of weight and leverage and know the secrets of the people who built the pyramids"; but when he died in 1951, Ed took his secrets about building sophisticated megaliths with him. How did he do it?

D.B. Cooper

On November 24, 1971, a passenger (wearing a black suit and listed as "Dan Cooper") boarded Northwest Orient's Flight 305 from Portland to Seattle. Once in the air, he threatened to detonate an explosive device and demanded $200,000 in $20 bills. He rather calmly and courteously told the pilot to land as planned in Seattle, where he released the passengers, picked up the money as well as four requested parachutes, and directed the Boeing 727's crew to fly to Mexico.

Flight 305 landed in Reno, but the mysterious hijacker was no longer aboard. At an altitude of 10,000 feet, he strapped the money onto his body, lowered the tail exit stairs, and parachuted out somewhere over rugged and mountainous terrain of Washington state. In 1980, a boy found $5,800 in $20 bills, with proper serial numbers, buried in a muddy bank along the Columbia River, but this is the only solid evidence recovered to date. If "D.B. Cooper" is alive, he is estimated to be in his mid- to late 80s. What do you think?

Lost Dutchman Mine

Even though the area east of Phoenix is called "The Superstition Mountains," it is actually a rugged, largely unexplored land of ravines, plateaus, and mesas that has seen more than its share of brutal deaths and mysterious vanishings. When the Conquistador Francisco de Coronado arrived there in 1540, while seeking the legendary "Seven Golden Cities of Cibola," Native Americans told him the area held much gold, but refused to help him because they feared that the "Thunder God" would kill them if they trespassed on his sacred land. After the Spaniards began to explore it for themselves, their men started to mysteriously vanish, and the bodies were found mutilated with their heads cut off. This terrified them so much that Coronado named this evil place *Monte Superstition*.

Don Miguel Peralta, a prominent rancher from Sonora, Mexico found a rich vein there in 1845, and over the next several years, his "Sombrero Mine" produced millions in gold. But by 1848, the Apaches had become so hostile he decided to halt mining and return to Mexico until things cooled off a bit. He carefully hid the mine's entrance, wiped out any traces of his activities, and packed the already-mined gold onto burros. He planned to leave and return later, but that did not happen.

Early the next morning, Indians massacred all the Mexicans and scattered their pack animals, spilling gold as they plunged over the cliffs and into deep ravines. For years, soldiers and prospectors found skeletons of burros with rotted leather packs, brimming with gold ore. In 1870, a doctor who had treated the Indians was blindfolded by tribal elders and taken into the area where he discovered a stack of gold nuggets piled near the base of a canyon wall.

The mine's current name comes from Jacob Walz, who was actually German, not Dutch. After searching for gold in North Carolina, Mississippi, California, and Nevada, "Snowbeard" (as the Indians called him) set out in 1870 with a Jacob Weiser into the lands around Superstition Mountain. Soon afterward, they were seen in Phoenix paying for supplies and drinks with gold nuggets. The pair did this for several decades, then Weiser vanished without a trace. Walz continued going back to the secret mine, returning with saddlebags full of gold. He died in 1891, with a sack of rich ore under his bed, but The Lost Dutchman Mine hadn't claimed its last victim yet.

Elisha Reavis was known as "The Madman of the Superstitions." He lived in the area, but the Apaches never bothered him because they were in awe as he ran naked through the canyons and fired his pistols at the stars. In 1896, one of his friends got concerned after Elisha was overdue for his periodic trip into town. His body had been eaten by the coyotes, but his head had been cut off (just like the Conquistadors) and it was found several feet away. A couple of Easterners went searching for the mine that same year and were never seen again. Is there a curse . . . or an angry Thunder God?

Money Pit

This pertains to one of the world's longest running quests for buried treasure. It started in 1795, when sixteen-year-old

Daniel McGinnis came to Nova Scotia's Oak Island on a fishing trip. Beneath the scarred branches of an old oak tree, he noticed a depression in the ground, roughly sixteen feet in diameter. With visions about pirate treasure swirling through his head, he returned the next day with two pals and picks and shovels, but this would be a much bigger job than expected.

About four feet down they hit a layer of flagstones, and after it was removed they encountered packed logs at ten, twenty, and thirty feet. They soon realized they needed better tools and returned to the mainland. Nine years would pass before they came back to the island, this time with financial backing and a force of local labor. As their digging continued, they hit oak platforms caulked with putty and coconut fibers at forty, fifty, and sixty feet down; a plain oak one at seventy feet, and another sealed with putty at a depth of about eighty feet. Ninety feet down, they uncovered a non-native stone bearing a rather cryptic inscription that was translated to say "Forty Feet Below Two Million Pounds are Buried."

They hastily removed the stone, only to expose another layer of wood. But it was getting quite dark by then, and water seepage was becoming a problem, so they decided to come back when it was daylight. The following day was Sunday, and when they returned early on that Monday morning they found the shaft filled with water up to about thirty feet from its rim. Their attempts to pump or drain the pit failed and it kept refilling, so they abandoned the project and returned home.

The following spring they dug a parallel shaft hoping it would drain off the water, and at roughly 110 feet down they cut into the original one. The walls of the new shaft collapsed, and the original pit filled back up with water to its earlier level.
What do you think lies at the bottom?

Voynich Manuscript

It has been called "the world's most mysterious manuscript" even though many people have never even heard of it. Named after the antique bookseller Wilfrid Voynich who acquired it back in 1912, it is a extremely detailed 240-page book written in a language or a script that is completely unknown. Its pages are filled with odd diagrams of strange events and colorful drawings of plants that do not appear to match any known species. While its author remains unknown, carbon dating indicated its pages were made sometime between 1404 and 1438AD.

Some have suggested that it describes medicinal preparations and drugs used in medieval or early modern medicine, while others claim the pictures of herbs and plants hint it might have been a sort of textbook for alchemists. Its unidentifiable biological drawings have led some to theorize it could have an alien origin; but considering the amount of time and effort required to create it, most agree it is very unlikely to be a hoax. What is it?

It's Up To You Now

You clearly have curiosity, courage, and commitment to have made it this far. All you need to do now is put a few simple clues together, but you better get started soon because the average life span of a civilization is only about 350 to 400 years, and the clock is already ticking.

But aren't you starting to wonder what the next Chapter is about? I know I am!

CHAPTER NINE:
FITNESS TEST

The Introduction claimed that this is a very different kind of book. If you still are not convinced, this final Chapter should do it. The following fitness test is provided to determine whether you slacked off during your mental exercise program, and it includes provocative questions on earlier sections.

Get a pencil and a separate sheet of paper upon which to record your answers to the following simple questions. So, are you ready to begin?

1. The visible light color with the shortest wavelength is:
    a. Red
    b. Violet
    c. Green
    d. Yellow

2. What color skin do Polar Bears have?
    a. Pink
    b. White
    c. Black
    d. None of the Above

3. Hypotheses are based on independent experimentation.
    a. True
    b. False

4. Which of the following involves bobsled tracks?
   a. Trugo
   b. Pesäpallo
   c. Wok Racing
   d. None of the Above

5. The Mai Tai was actually invented in Tahiti.
   a. True
   b. False

6. Which of the following is defined as a global characteristic?
   a. Climate
   b. Weather
   c. Both of the Above
   d. None of the Above

7. Which of the following typically have the highest blood pressure?
   a. Mice
   b. Giraffes
   c. Humans
   d. Armadillos

8. Earth's current atmosphere consists of approximately 78 percent:
   a. Carbon Dioxide
   b. Water Vapor
   c. Nitrogen
   d. Oxygen

9. On what day *of the week* did Dutch explorer Roggeveen discover Rapa Nui?
   a. Friday
   b. Sunday
   c. Monday
   d. Tuesday

10. According to the Big Bang Theory, the Universe is approximately:
    a. 13.75 Billion Years Old
    b. 4.5 Billion Years Old
    c. 65 Million Years Old
    d. 6,000 Years Old

11. Which of the following actresses replaced all of the toilet seats on her estate whenever she re-married?
    a. Bette Davis
    b. Joan Crawford
    c. Marilyn Monroe
    d. Elizabeth Taylor

12. In which of the following states is it specifically illegal to hunt camels?
    a. Oregon
    b. Indiana
    c. Arizona
    d. Mississippi

13. Which of the following originated as a survival skill for runaway slaves?
    a. Buzkashi
    b. Capoeira
    c. Nalakatuk
    d. Sepak Takraw

14. Which of the following is often found frozen beneath ocean sediments?
    a. Ethanol
    b. Methanol
    c. Methane Hydrate
    d. Non-Fossil Biogas

15. What is the maximum age for becoming a pilot in Turkey?
    a. Fifty
    b. Sixty
    c. Seventy
    d. Eighty

16. Bias is simply another term for prejudice.
    a. True
    b. False

17. Which of the following is/are generally not considered as criteria for being "civilized"?
    a. Dense Urban Settlements
    b. A Written Language
    c. Neither a nor b
    d. Both a and b

18. Which of the following sports is played with a bat and ball?
    a. Ga-Ga
    b. Korfball
    c. Pesäpallo
    d. Bossaballl

19. The so-called *"Big Whack Theory" concerns*:
    a. A Principle of Organized Crime
    b. The Creation of the Universe
    c. The Formation of the Moon
    d. None of the Above

20. Which of the following ancient civilizations did not arise in Mesopotamia?
    a. The Caral
    b. The Valdivians
    c. The Mycenaeans
    d. All of the Above

21. Which of the following animals can get malaria and carry leprosy?
    a. Mosquitoes
    b. Armadillos
    c. Crocodiles
    d. Moles

22. Galileo was punished for suggesting that the Earth was the center of the Universe.
    a. True
    b. False

23. The diesel engine demonstrated at the 1900 World's Fair in Paris ran on:
    a. Diesel Oil
    b. Hydrogen
    c. Peanut Oil
    d. Gunpowder

24. Which of the following sports originated in England?
    a. Cheese Rolling
    b. Extreme Ironing
    c. Bog Snorkeling
    d. Both a and b

25. Which of the following is not considered to be a principal greenhouse gas?
    a. Ozone
    b. Oxygen
    c. Methane
    d. Water Vapor

26. Who does the *King of Clubs represent* in a standard deck of playing cards?
    a. Julius Caesar
    b. Charlemagne
    c. Biblical King David
    d. Alexander the Great

27. Most commercial Hydrogen is currently produced from fossil fuel.
    a. True
    b. False

28. The competitive sport of *Wife-Carrying* was invented in:
    a. Finland
    b. England
    c. Belgium
    d. Australia

29. Which of the following ancient cultures had a special fascination with bulls?
    a. The Aztecs
    b. The Minoans
    c. The Phoenicians
    d. The Mycenaeans

30. Which of the following can be a principal trigger for major temperature changes?
    a. Continental Movements
    b. Ocean Current Changes
    c. Both a and b
    d. None of the Above

31. Roughly what percentage of a barrel of crude oil is typically converted into transportation fuels?
    a. 20 to 25 Percent
    b. 40 to 50 Percent
    c. 70 to 75 Percent
    d. Over 85 Percent

32. Which of the following ancient civilizations originated in North America?
    a. The Incas
    b. The Nazcas
    c. The Olmecs
    d. The Anasazis

33. What was President Harry Truman's middle name?
    a. Sidney
    b. Stanley
    c. Sherman
    d. None of These

34. The basic composition of the Earth's atmosphere hasn't changed much over time.
    a. True
    b. False

35. Which of the following Greeks wrote about the lost continent of Atlantis?
    a. Plato
    b. Aristotle
    c. Archimedes
    d. None of Them

36. It is possible to produce both gas and liquid fuel from coal.
    a. True
    b. False

37. Rolling down hillsides inside large spheres is called:
    a. Ga-Ga
    b. Zorbing
    c. Korfball
    d. Bossaball

38. Which of the following is officially defined as a long-term statistical value?
    a. Weather
    b. Climate

39. The Feathered Serpent God of several Mesoamerican cultures was called:
    a. Kukulcán
    b. Virakocha
    c. Quetzalcoatl
    d. a and c, but not b

40. Soot produced by burning biomass and some fossil fuels:
   a. Decreases the Surface Temperature
   b. Increases the Surface Temperature
   c. Both a and b Above
   d. None of the Above

41. Which of the following cultures left a terra cotta army?
   a. The Ancient Chinese
   b. The Ancient Greeks
   c. The Ancient Romans
   d. The Ancient Egyptians

42. Which of the following sports combines tag and holding one's breath?
   a. Kabaddi
   b. Pesäpallo
   c. Nalakatuk
   d. Sepak Takraw

43. Winston Churchill was reportedly born:
   a. At a Party
   b. In a Palace
   c. Prematurely
   d. All of the Above

44. Which of the following is/are characteristic(s) of a "fuel'?
   a. Relatively Stable
   b. Easily Transportable
   c. Able to Store Energy
   d. All of the Above

45. The most popular theory suggests that humans first migrated to North America from:
   a. Asia
   b. Africa
   c. Europe
   d. Polynesia

46. In Thailand, it's against the law to leave your house without any:
    a. Shoes
    b. Money
    c. Underwear
    d. All of the Above

47. Which of the following fuels can be made from switch grass?
    a. Ethanol
    b. Biomass
    c. Neither a nor b
    d. Both a and b

48. Which of the following built earthen mounds in North America?
    a. The Anasazis
    b. The Zapotecs
    c. The Valdivians
    d. The Mississippians

49. Which of the following was used to depict blood in the movie "Psycho"?
    a. Catsup
    b. Grape Juice
    c. Chocolate Syrup
    d. None of the Above

50. Under USLMRA Rule, what is the minimum age for race competitors?
    a. Twenty-One
    b. Eighteen
    c. Twelve
    d. Eight

Time's up! Put down your pencil, and grade your answers by comparing them with the following:

1. b. Violet light has the shortest wavelength of these while red light has the longest.

2. c. Polar bears actually have black skin, and the outer guard hairs of their fur are translucent.

3. b. By their official definitions, hypotheses are based upon observation, while scientific theories must be supported by independent experimentation.

4. c. And it's done by skidding down icy ditches in Chinese cookware.

5. b. The most popular legend says it was actually invented in Oakland, California.

6. d. According to their official WMO definitions, neither are considered to be global phenomena.

7. b. An adult giraffe's heart can also weigh up to twenty-five pounds.

8. c. And along with Oxygen, it makes up most of the Earth's atmosphere.

9. b. It was on Easter Sunday, which is why he called it Easter Island.

10. a. But you should probably recognize these other dates as well.

11. b. Joanie kept all those plumbers busy in Brentwood.

12. c. This may be why Arizona has an illegal camel immigration problem.

13. b. So try to avoid getting into the middle of a circle of Brazilians.

14. c. But be careful when drilling into Methane Hydrate deposits!

15. d. Apparently there actually *are* some old pilots in Turkey!

16. b. According to its official definition, Bias can sometimes be based upon reasoned judgment and can help us make better decisions.

17. c. Several great ancient civilizations have not exhibited these.

18. c. All you have to do is remember a few simple game rules.

19. c. Don't confuse this with theories about Big Bangs or Big Macs.

20. d. The Caral and Valdivians were in South America, and Mycenaeans were in Greece.

21. b. It's the price they pay for being for so darn cute and cuddly.

22. b. Never get your Galileo's confused with your Ptolemy's.

23. c. Squeeze more of those goober peas, s'il vous plait!

24. d. Bog Snorkeling was actually invented in Wales.

25. b. But if it was, we would all have to inhale more.

26. d. The others are represented by the other suits.

27. a. Most of it comes from natural gas, a fossil fuel.

28. a. Those same frisky Finns also gave us Pesäpallo.

29. b. They even kept a Minotaur in the basement.

30. c. Because both can affect distribution of heat energy.

31. c. This includes gasoline, diesel oil, etc.

32. d. And they lived in the Four Corners area of the American Southwest.

33. d. It doesn't stand for anything, but was a compromise to satisfy both his grandparents, whose last names were Shippe and Solomon.

34. b. Scientists say the Earth has had at least three different atmospheres.

35. a. Plato related the story of Atlantis in his Socratic dialogue *Timaeus*.

36. a. It's been done a number of times already.

37. b. So hop into those zorbs, and roll on Dude!

38. b. But weather is real stuff happening in real time.

39. d. Virakocha was a different character in South America.

40. c. If suspended, soot can cool the underlying surface; but if deposited on an ice field or glacier, soot can warm the surface.

41. a. It was found in the tomb of the Qin Dynasty's first Emperor.

42. a. And it's quite popular among schoolchildren in India.

43. d. Pop another bottle of bubbly, Winnie's here!

44. d. Remember that fuel is not the same thing as energy.

45. a. Although less popular theories suggest other routes.

46. c. Does this mean kilt-wearing Scotsmen have to stay inside?

47. d. And it doesn't affect the price of crops like corn or soybeans.

48. d. One of their largest ones is at Cahokia in southern Illinois.

49. c. And I've heard that you can also use it on ice cream.

50. d. But only with their parent's permission, of course.

How did you do? If you did not get at least thirty-five answers (70 percent) correct, you need to re-read this book because apparently your brain could use a little more exercise.

CONCLUSION

I hope you enjoyed my book, and perhaps learned something along the way. It is quite different from my other works, which include: textbooks, technical manuals, a nautical dictionary, collections of short stories, an environmental handbook, a cookbook, a novel, and more than one hundred feature articles for magazines and Web sites.

Some of the seemingly silly sections may have looked like mental meandering, but they were actually part of a subtle strategy to exercise the skills required to survive in modern society. The *Warm Up Exercises* were designed to stimulate your curiosity about unfamiliar topics, while *Chapter One* laid a foundation for clear thinking by defining the key differences between rumors and realities.

Chapter Two next exercised your courage to learn more about one of today's most controversial topics, and revealed how self-centered greed and naïve ignorance have affected public perceptions, politics, and poorly-informed decisions about climate change. It is followed by a "cool down" session in *Chapter Three*, which is actually a tongue-in-cheek way to exercise another important modern survival skill, communicating effectively.

Chapter Four exercised your curiosity, courage, and commitment to discover more about our less famous, but fascinating, ancestors; and offered valuable lessons in human history. The following "cool down" session in *Chapter Five* examined some weird ways that folks recreate; while *Chapter Six* explored the more challenging and controversial topic of alternative energies and fuels, and described how greed, ignorance, and apathy hamper development of a viable energy Plan.

Chapter Seven may have appeared irrelevant, but it actually proved how these three threats, if left unchecked, result in ridiculous legislation. *Chapter Eight* provided a workout plan that challenged your brain with some of history's mysteries, and the fitness test in *Chapter Nine* determined whether you fell asleep along the way.

You have now completed the premier mental exercise program for flabby modern minds. So get out there and: hypothesize, theorize, complain about the weather (or the climate), start witty conversations, discuss archeology, play lesser-known sports, alternate energies and fuels, cite silly laws, and solve history's mysteries. But most importantly, keep thinking!

Roger Huff

RESOURCES & REFERENCES

Publications

- Heyerdahl, Thor, Daniel Sandweisss, & Alfredo Narvaez. 1995. *Pyramids of Tucume: The Quest for Peru's Forgotten City.* London: Thames & Hudson.

- Huff, Roger Paul. 2006. *Captain Bucko's Water & Weather Handbook: An Entertaining and Easy-To-Read Collection of Inside Information, Fascinating Facts, Trivial Tidbits, and Helpful Hints by a Professional Oceanographer and Marine Meteorologist to Help Make Your Voyages Safer and more Enjoyable.* iUniverse

- Huff, Roger. 2011. *Journey of the Lost Princess: Adventure and Romance in the Mysterious Land of the Incas.* iUniverse

- Longhena, Maria & Walter Alva. 1999. *The Incas: And Other Ancient Andean Civilizations.* Barnes & Noble Books

- 2006. *Insight Guides: Peru*, APA Publishing

Web Pages

http://www.ancient-wisdom.co.uk/top50stones.htm

http://www.archaeologywordsmith.com/lookup.php?category=&where=headword&terms=prehistory

http://climate.nasa.gov/

http://climate.nasa.gov/

http://www1.eere.energy.gov/geothermal/future_geothermal.html

http://en.wikipedia.org/wiki/Alternative_energy

http://en.wikipedia.org/wiki/Alternative_fuel

http://en.wikipedia.org/wiki/Ancient_Egypt

http://en.wikipedia.org/wiki/Ancient_Greece

http://en.wikipedia.org/wiki/New_World

http://en.wikipedia.org/wiki/Ancient_Pueblo_Peoples

http://en.wikipedia.org/wiki/Ancient_Rome

http://en.wikipedia.org/wiki/Aztec

http://en.wikipedia.org/wiki/Big_Bang

http://en.wikipedia.org/wiki/Cradle_of_civilization

http://en.wikipedia.org/wiki/Dumb_laws

http://en.wikipedia.org/wiki/Easter_Island

http://en.wikipedia.org/wiki/Electric_power_transmission

http://en.wikipedia.org/wiki/Flood_myth

http://en.wikipedia.org/wiki/History_of_the_Earth

http://en.wikipedia.org/wiki/Inca_Empire

http://en.wikipedia.org/wiki/Indus_Valley_Civilization

http://en.wikipedia.org/wiki/Maya_civilization

http://en.wikipedia.org/wiki/Mesopotamia

http://en.wikipedia.org/wiki/Minoan_civilization

http://en.wikipedia.org/wiki/Mississippian_culture

http://en.wikipedia.org/wiki/Mycenaean_Greece

http://en.wikipedia.org/wiki/Noah's_Ark

http://en.wikipedia.org/wiki/Norte_Chico_civilization

http://en.wikipedia.org/wiki/Olmec

http://en.wikipedia.org/wiki/Phoenicia

http://en.wikipedia.org/wiki/Quetzalcoatl

http://en.wikipedia.org/wiki/Teotihuacan

http://en.wikipedia.org/wiki/Trivia

http://en.wikipedia.org/wiki/Valdivia_culture

http://en.wikipedia.org/wiki/White_Gods

http://en.wikipedia.org/wiki/Zapotec_civilization

http://guyism.com/sports/10-strange-but-real-sports-you-havent-heard-of.html

http://inventorspot.com/articles/10_weird_sports_from_
around_the_world_15185

http://nsidc.org/arcticmet/basics/weather_vs_climate.html

http://science.nationalgeographic.com/science/space/
universe/origins-universe-article/

http://solarsystem.nasa.gov/scitech/display.cfm?ST_ID=446

http://voices.yahoo.com/the-mayan-mystery-was-kukulcan-
quetzalcoatl-305566.html?cat=37

http://www.afdc.energy.gov/afdc/

http://www.altenergy.org/

http://www.alternative-energy-news.info/

http://www.altfuels.org/backgrnd/altftype.html

http://www.beyondfossilfuel.com/alternative_fuels.html

http://www.bibliotecapleyades.net/profecias/esp_
profecia01h1.htm

http://www.biodieselsustainability.com/feedstocks/

http://www.brainyquote.com

http://dornsife.usc.edu/news/stories/804/methanol-
a-fuel-for-the-future/

http://www.dumblaws.com/

http://www.epa.gov/climatechange/

http://www.epa.gov/climatechange/glossary.html

http://www.godandscience.org/evolution/lifeonearth.html

http://www.godandscience.org/evolution/lifeonearth.html

http://www.idiotlaws.com/

http://www.infoplease.com/spot/prestrivia1.html

http://www.israelect.com/reference/WesleyASwift/extra/ WHITEGOD.htm

http://www.lawguru.com/weird/part01.htm

http://www.lawguru.com/weird/part02.html

http://www.merriam-webster.com/dictionary/bias

http://www.merriam-webster.com/dictionary/fact

http://www.merriam-webster.com/dictionary/hypothesis

http://www.merriam-webster.com/dictionary/mythology

http://www.merriam-webster.com/dictionary/opinion

http://www.merriam-webster.com/dictionary/rumor

http://www.merriam-webster.com/dictionary/theory

http://www.nasa.gov/mission_pages/noaa-n/climate/ climate_weather.html

http://www.newscientist.com/article/dn17240-methanol- challenges-hydrogen-to-be-fuel-of-the-future.html

http://www.nwcreation.net/noahlegends.html

http://www.nws.noaa.gov/om/brochures/climate/
Climatechange.pdf

http://www.pbs.org/wgbh/nova/space/history-universe.html

http://www.pbs.org/wgbh/nova/space/origins-series-
overview.html

http://phys.org/news/2011-12-methanol-hydrogen-gas-fuel-
future.html

http://www.sciencedaily.com/news/matter_energy/
alternative_fuels/

http://www.scientificamerican.com/article.cfm?id=the-
truth-about-fracking

http://www.scientificamerican.com/article.cfm?id=future-
of-clean-coal-tied-to-success-of-carbon-capture-and-
storage

http://www.strangefacts.com/laws.html

http://www.thebioenergysite.com/articles/15/biomass-
technologies-of-the-future

http://www.topendsports.com/sport/unusual/index.htm

http://www.travelandleisure.com/articles/worlds-strangest-
sports

http://www.usefultrivia.com/

http://www.wmo.int/pages/prog/arep/gaw/ghg/ghgbull06_
en.html

http://www.wmo.int/pages/prog/gcos/documents/ gruanmanuals/GAW/QC-Related%20Terminology.pdf

http://www.yosemite.epa.gov/r10/AIRPAGE.NSF/webpage/ Biodiesel+-+A+Fuel+For+the+Future

Front Cover Image:

http://upload.wikimedia.org/wikipedia/commons/5/5e/BH_ LMC.png (Credit: Alain Riazuelo)